W9-BNL-554

ANTS AT WORK

. . . .

How an Insect Society Is Organized

•

Deborah M. Gordon

. . . .

Illustrations by

Michelle Schwengel

THE FREE PRESS

THE FREE PRESS
A Division of Simon & Schuster Inc.
1230 Avenue of the Americas
New York, NY 10020

Designed by Pei Koay

Manufactured in the United States of America

10 9 8 7 6 5 4 3 2 1

Library of Congress Cataloging-in-Publication Data
Gordon, Deborah (Deborah M.)
Ants at work : how an insect society is organized / Deborah Gordon.
p. cm.
Includes bibliographical references.
1. Ants—Behavior. 2. Insect societies. I. Title.
QL568.F7G64 1999 99-35853 CIP
595.79′6—dc21

ISBN 0-684-85733-2

ACKNOWLEDGMENTS

Many people, more than I can list, have helped with the research I describe in this book. My greatest debts are to all of my teachers, especially my mother, Barbara Gordon; my dissertation advisor, John Gregg; and Richard Lewontin. I would also like to thank all of the people who have at different times provided the places or means for me to work, especially my father, Jack Gordon, and Stephen Wainwright, William Hamilton, and John Lawton. A host of students have helped with field and laboratory work, and I am grateful to them all: M. Allinei, J. Augustine, B. Bailey, M. Baray, K. Barton, M. Bateson, V. Braithwaite, M. Brown, D. Cotter, B. Cuevas, A. Dalrymple, R. Daniel, S. DePue, C. Desjardins, P. Dohrenwend, S. Dominick, S. Elder, D. Ferrier, K. Filseth, N. Foerstoefl, M. Forgy, S. Foote, A. Fullerton, B. Gavino, H. Graham, M. Hagerty, R. Hallett, T. Hill, K. Human, R. Hung, C. Johnson, S. Josephson, A. Kay, T. Kennedy, L. Kiriakos, E. Lewis, A. Liu, H. Liu, H. Luessow, C. Mueller, R. Paul, B. Raymond, K. Roth, M. Rollins, T. Rucker, N. Sanders, E. Saxon, E. Marin-Spiotta, G. Tarling, F. Weigand, M. Weiser, and P. Wright. I thank Finn Richards, and more recently Ned and Annabel Hall, for the use of their land as a study site.

This book was written during an idyllic semester at the Stanford Center for Advanced Studies in Behavioral Sciences; my fellowship there was supported in part by NSF SBR 9601236. Kathleen Much, the editor at the Center; Samuel Beshers; Henry Horn; Ross Gelbspan; and especially John Gregg all did a great deal to make the original manuscript into this book, and it has been much improved by Stephen Morrow, my editor at The Free Press. My agent, John Thornton, has been enormously helpful.

Finally, my thanks to Ben for his unwavering interest in ants.

CONTENTS

INTRODUCTION

•

The basic mystery about ant colonies is that there is no management. A functioning organization with no one in charge is so unlike the way humans operate as to be virtually inconceivable. There is no central control. No insect issues commands to another or instructs it to do things in a certain way. No individual is aware of what must be done to complete any colony task. Each ant scratches and prods its way through the tiny world of its immediate surroundings. Ants meet each other, separate, go about their business. Somehow these small events create a pattern that drives the coordinated behavior of colonies.

I study ants because I am interested in linking levels of organization. The living world is structured in layers, from molecules to cells to organs to individuals to populations to ecosystems. A fundamental question in biology is how events at these different levels are related. A thought happens when electrical impulses move around the tangle of neurons in a brain, but a thought is something more than, and something other than, neurons. I set out to do research on the connections between the tiers of natural organization, and I chose to work with ants because they invite the observer to notice three distinct levels: ants, colonies, and populations. All ants live in colonies, consisting of sterile workers and one or more queens, who produce the workers and the next generation of reproductives, new queens and males. The new reproductives fly off to mate and then begin new colonies. The population is all of the colonies whose reproductives could mate with each other.

In studying ant behavior, I try to figure out how each ant decides what to do, and how this adds up to the achievements of colonies. What fascinates me about ants is that this back-and-forth, from ants to colonies and colonies to ants, is always visible and impossible to forget. Because ants are separate beings that move around freely, they attract attention as individuals. But nothing ants do makes sense except in the context of the colony. Zoom in, you see ants—zoom out, you see a colony. Ants and colonies are both there in front of you, all the time.

Ant-colony life has evolved over the past 100 million years from the less social lives of the ants' ancestors, the wasps. The evolution of colony behavior hinges on how well a colony functions, relative to the other colonies in its population. If a colony that acts a certain way tends to produce more offspring, and variation in colony behavior is heritable, then over many generations the more successful type will be better represented in the population than less successful ones. In fact, we know nothing about the inheritance of ant-colony behavior, and little about the conditions in which the behavior that we now see once evolved. As a kind of shortcut over these chasms of ignorance, we can pretend that success in the ant world is the production of many new colonies. If evolution is happening now, we can see it only by asking ecological questions, about why some colonies might reproduce more than others. All of a colony's behavior, how it gets resources and how it uses them, contributes to its reproduction. To see evolution in action, we must look at the colony's livelihood: where it lives and what it eats and who eats it and with whom it shares its food and space.

For the past seventeen summers I have studied the harvester ants in a small patch of the Arizona desert. My approach has been: Find a pattern, and then figure out how and when it changes. I ask how the behavior of an ant fits into the life of its colony, and how in turn the relations of neighboring colonies may shape, over evolutionary time, the behavior of ants within colonies. Because I kept track of the same colonies, about 300 of them, for many years, I followed colonies through their life cycles. New colonies are born, struggle to occupy a foraging area, grow larger, start to reproduce, and then settle in among their lifelong neighbors. I discovered that how colonies behave and how they relate to their neighbors both change as the colony grows older and larger. Neighboring colonies search the same places for food, so pressure from neighbors can starve out a tiny new colony or keep an older one from reproducing. Colonies live for fifteen years, but each ant lives only a year. When an older colony acts different from a younger one, it is not following the advice of older, more experienced ants. This leads me to ask why, if the ants in young colonies behave just like ants in old ones, the colonies seem to become more staid and prudent as they get older and larger.

It is difficult to think about how an ant colony works. Not only is an ant colony's behavior complex, woven from zillions of ant acts, but all those tiny events add up to something different from any society we know. Stories about totalitarian societies, inexorable armies, and voracious monsters are often told as stories about ants. But ants have no dictators, no generals, no evil masterminds. In fact, there are no leaders at all. The relations between neighboring colonies are as intriguing as the inner workings of a single ant society. Ecological accounts of animals' lives often portray a kind of suburban nightmare, each individual or family struggling furiously to accumulate more and better resources than its neighbors, with the richest one the evolutionary winner. But the ants have no property, no bank accounts, no picket fences. In a harvester ant neighborhood, there are no borders. This is the puzzle. If the ants don't work like a miniature human society, how do a group of rather inept little creatures create a colony that gets things done?

1

THE RHYTHMS OF THE LANDSCAPE

•

Red ant or ant of red abdomen

It is somewhat average in size, a little firm, a little hard, ruddy. It has a heap of sand, a mound of sand, a hill. It sweeps, makes itself sand heaps, makes wide roads, makes itself a home. It is the worst one to bite, if it bites the foot [the effect] extends to the groin; if it bites the hand, it extends to the armpit; it swells.

The Florentine Codex: Fray Bernardino de Sahagun,
General History of the Things of New Spain, 1590.
[Translation of Aztec description of the red harvester ant.]

•

25 ACRES

I study the ants at the side of a rough paved road that runs through a flat valley between the Chiricahua and Peloncillo mountains at the state line of Arizona and New Mexico. An enormous sky surrounds an endless reach of land. The Chiricahuas, to the west, are so close one can see the patches of rock change color during the day. The Peloncillos, to the east and north, form a jagged outline in the distance. To the south the desert stretches across the eighty miles to Mexico.

I stay at the Southwestern Research Station up in the Chiricahua mountains. The station belongs to the American

Museum of Natural History in New York. At the peak of the summer season there might be fifty people staying there, mostly undergraduates who come to work either for the station or as research assistants for people like me. We get up at 4:30, when it is still dark, and meet in the dining room, where I usually wake myself up by industrious chewing of my Grape-Nuts. Then there is some antlike milling around while people make their peanut-butter-and-jelly sandwiches to get them through the morning, collect their water bottles and clipboards, and pile into the van. It's about a twenty-minute drive down into the desert. When I first came to work at SWRS, as a graduate student and then as a postdoctoral researcher, I rented cars, eventually moving from the aptly named Rent-a-Wreck to small rental cars that look embarrassed when covered with dust and turn out to have spare tires the size of a doughnut. But when I joined the faculty at Stanford, the university bought a huge air-conditioned van for the purposes of ant research, so we bump down the dirt road from the station in great style. Along the road I can usually avoid the rabbits that seem to throw themselves at the wheels of our van. Down in the flats, as I turn the corner at the Arizona–New Mexico state line, I adjust the visor to block the sharp glow of the rising sun.

You can recognize an ant researcher by looking at her ankles: We wear our socks over the cuffs of our pants. Experience with stinging ants teaches that you can see them on your hands, and feel them go down the back of your neck, but they can get up inside your pants faster than you can brush them off. However, in the desert, keeping off the sun is more difficult than keeping off the ants. Some people wear shorts, at least until they have been stung. Over the years I have evolved a costume that includes a long-sleeved shirt, a cap with a kind of curtain around its lower edge, and the largest sunglasses I can find. I look rather like an insect myself.

For the first few years my assistant and guide was Kristin Roth, who grew up at the station while her father was its director. Kristin taught me to distinguish landmarks in the desert. At first everything looked alike, and I could walk a hundred yards and have no idea how far I'd come. Kristin would stride past me in her cowboy boots, saying "Colony 81? It's right by that little bush over there." Since then, one to five students each year have come out to Arizona to work with me. Some love the desert; one year they happily played Frisbee with cow pies. Some are fascinated by ecology, enough to go on to graduate school. A few have been hooked for life on ants. Unfortunately, some discover that waking up at 4:30 A.M. to follow ants around and slowly roast is not their idea of fun. I do my best to warn everyone ahead of time that ant-watching in the desert is hard work, so anyone brave enough to come with me in the first place tends to stick it out.

The 25 acres I have inspected inch by inch, for the past seventeen summers, is part of a 7,000-acre cattle ranch. When I first saw the site in July 1981, it looked like an Alpine meadow, filled with a stunning variety of flowers. The land was owned by a rancher who fenced off areas of a few hundred acres each and rotated the cattle from one paddock to another. The profusion of flowers was the result of a few wet summers and a few years without cattle. In the late eighties the land was sold to another rancher who took down all the fences and stopped rotating the cattle. Because of heavy grazing and some dry summers, the site now looks very different from the flowery meadow of 1981. Much of the ground is bare, and only the toughest shrubs that cattle cannot eat survive: snakeweed (*Guttieriza*), Mormon tea (*Ephedra*), and acacia.

It rains in the spring and then more fiercely in July and August, when monsoon storms bring a few weeks of heavy rains, flash floods, and then, overnight, the first sprouts of green grass. In 1996 large areas of the site were underwater for a week after the heaviest rain, and then came a carpet of green grass. Some of the flowers returned, annuals whose seeds had waited out the many dry years since the early eighties.

On the first day there every summer, I park the van, get out, and walk on to the site with great relief and exhilaration; relief because once again the ants are still there—no one has bought the land and built a convenience store on it; exhilaration at the prospect of shifting down from the human to the ant scale, from large to small and from the slow, emphatic tuba notes of human affairs to the light harpsichord trills of ant life. The ants are so busy; their business is so complicated; and on this little patch of desert they play out a perpetual rich drama, oblivious to all the other dramas that occupied me for the rest of the year. I discover again my irrelevance to the ants.

THE ANTS

The ants begin to come out at sunrise and stay out until about 11, when it gets so hot that they go back into their nests. Early in the morning we hear the calls of coyotes and the songs of desert larks. We see the occasional tarantula out for a morning walk along the warm asphalt road and, rarely, a rattlesnake too full and lazy to go back into its nest after a large nocturnal meal. Here and there is a collapsed kangaroo-rat mound; a few contain live rats, and recently ground squirrels have moved into some of the abandoned mounds. After rain the dung beetles earnestly roll their balls of cow dung much larger than themselves. Shiny black bombardier beetles point their rear ends in the air and threaten to spray acetic acid. The harlequin grasshopper has a red, blue, yellow, and green costume so much more like the cover of the *Sgt. Pepper's Lonely Hearts Club Band* album than real life that when I first saw it I did not believe my eyes. The whiptail (*Cnemidophorous*) lizards undulate from the cover of one bush to another, offering brief flashes of their blue bellies, working up to the frenzy of the midday heat when they rush about so fast their tails don't touch the ground. Sometimes a desert tortoise walks by. For several summers an old jackrabbit came out from behind the same bush every few days to watch me, but he or she eventually disappeared, and the ones I have seen since are more reserved.

On the human scale, the site looks like a plain of chaparral scrub with mountains on either side. On the ant-colony scale, the site looks like a bumpy, sandy surface with lots of gray, orange, and pink boulders and the occasional bush or plant. By the time it gets warm, about 7:30 in the morning, the ground quietly teems with harvester ants. They are called harvester ants because they eat mostly seeds, which they store inside their nests. They will happily take termites as well when they can find them.

There is a nest of the red harvester ant (*Pogonomyrmex barbatus*) every few yards. Young harvester ant colonies often start out with nests under bushes; eventually they destroy the bush roots as the nest expands underground, leaving an open space by the time the colony is four or five years old. An old, established colony may have a flat disk or mound a meter wide, covered with tiny pebbles, with one or two entrances in the middle. Colonies live for fifteen to twenty years, and most colonies stay in the same nest all their lives.

Harvester ants of this species are large (about 1 centimeter long) and brown. They have big heads that make them appear to jostle along as they walk. The back end does the propulsion and at every step pushes the heavy head forward with a little bump. The ants are sedate, even plodding. They spill out of the nest in the morning like elderly tourists pouring out by the busload. There are many species of *Pogonomyrmex* in the deserts of the southwestern United States and northern Mexico, and they are all known for their powerful sting, which they use to defend themselves against other ants and careless humans. Perhaps because their potent venom makes aggression unnecessary, they will rarely attack, and they tend to avoid confrontation.

As many as fifty other ant species may live on the site. The relations between red harvester ants and the other ant species sometimes get nasty, but never escalate into military raids. I have never seen a red harvester ant enter the nest of another species, except for foragers that soon hurry back out apparently embarrassed by their mistake; and I have never seen ants of any other colony, same or different species, raid a red harvester ant nest. Fighting does sometimes break out between colonies of red harvester ants, and we shall see how neighbors of this species work out their differences.

My view of the other ant species at the site is Pogo-centric, in that I think of them as belonging to three categories: other harvester ant species in the genus *Pogonomyrmex,* other big ants, and little ants. The differences in size may not matter as much to the ants as they do to me; my categories reflect how well I can see them. I think of an ant as a big one if it is large enough for me to see its separate parts and how they move; then it looks to me like an animal doing something. Ants that are smaller than about half a centimeter look to me like straight lines with antennae. Many species of tiny ants have been studied in the laboratory under a dissecting microscope, but I prefer to watch animals I can see without having to fiddle with any equipment. This might be why I do not study bird behavior, a field that requires skill with binoculars. It is certainly one of the reasons I chose to work on harvester ants in the desert: large brown ants on pale yellow sandy soil are easy to see.

Ants of the species I study, red harvester ants, are the largest of the *Pogonomyrmex* species at the site. Another species that like this one has large ants and large colonies, *P. rugosus,* tends not to overlap in range with *P. barbatus.* There is a patch of *P. rugosus* colonies on the other side of the road, and there was one colony in my study site for a few years, but it died. The other four *Pogonomyrmex* species at the site have ants smaller than those of *P. barbatus;* of these, the ants of *P. maricopa* are largest. They tend to be active later in the day than the ants of *P. barbatus,* as they can tolerate higher temperatures. *P. maricopa* colonies build crescent-shaped nests, and I think they may move their nests every few weeks during the summer rains. Two other species, *P. californicus* and *P. desertorum,* have much smaller workers and smaller colonies than *P. barbatus. P. desertorum* ants are rather timid. Ants of the last species, recently transferred to the genus *Ephebomyrmex,* are minuscule versions of *P. barbatus* ants, and live in tiny, practically invisible nests with strange workerlike queens. If the latter three species (*P. desertorum, P. californicus* and *E. imberbiculus*) did not have the special distinction of being in the genus *Pogonomyrmex,* I would group them with the little ants.

Of the other big ants, the species most important to the red harvester ants at this site is probably *Aphaenogaster cockerelli,* who eats seeds as the *Pogonomyrmex* do, but also eats insects. The *Aphaenogaster* are large, quick, graceful ants that can run circles around a red harvester ant worker, and often do. *A. cockerelli* are active at night and cannot tolerate high temperatures, so they interact directly with red harvester ants only early in the morning and on cool, overcast days. Sometimes ants of both species find an appealing dead insect, and then they will fight. A red harvester ant has its strong venom, while an *Aphaenogaster* has no sting at all, but the *Aphaenogaster* are so much quicker and more aggressive that they tend to win disputes. They also carry out the nefarious practice of blocking up the nest entrances of neighboring red harvester ant colonies at night. This means that only the late morning light reaches inside the red harvester ant nest, so the harvester ants come out later to forage. Because harvester ants must return to their nests once the midday sun raises the soil temperature above about 52 degrees C., a late start shortens their daily foraging period, leaving more food for the *Aphaenogaster* when they come out again that night.

The other big ants are the honeypot ants, *Myrmecocystus,* which eat mostly insects and other arthropods, and inside the nest store food as a sugary liquid inside the swollen abdomens of workers, thus earning the name "honeypots." One species, *M. mimicus,* is very common in some years, and in those years can be quite annoying to the harvester ants. The *Myrmecocystus* tolerate higher temperatures than red harvester ants so they tend to be active at midday, at the end of the red harvester ant foraging period. They come in waves and run around in circles on the harvester ant mounds, upsetting the foragers heading back into the nest with their loads. The honeypot ants may sometimes induce a harvester ant forager to drop its load, but more often the honeypot ants just seem to be making a nuisance of themselves for no immediate reward. A second honeypot species, *Myrmecocystus mexicanus,* a ghostly yellow nocturnal ant, is extremely rare at the site.

The abundant species in the "little ant" category include the small army ant, *Neivamyrmex nigrescens.* When on the move they raid the nests of other ants, spinning a branching network of curved trails across the desert, but they do not raid *Pogonomyrmex* nests, probably deterred by *Pogonomyrmex* venom. There are the little seed harvesters *Pheidole militicida* and *Pheidole desertorum;* the omnipresent *Forelius,* a tiny, pesky red ant; and *Solenopsis xyloni,* a close relation of the fire ant. The seed-eaters among these seem to collect seeds smaller than the ones used by the *Pogonomyrmex barbatus.* Occasionally a *Pheidole* colony builds a nest at the edge of a red harvester ant mound, and the red harvesters just ignore them. The *Pheidole* move quite frequently, and given the fifteen years a red harvester ant colony occupies its mound, maybe a brief incursion by a *Pheidole* colony is not worth bothering with.

The red harvester ants usually appear quite unconcerned with this tapestry of ant activity, and I will not discuss it further, except to say that we are now beginning to study some other ant species to find how they matter to red harvester ants.

Other colonies of the same species are, however, very important. Standing at a red harvester ant mound, looking around the neighborhood, you may see three or four other red harvester colonies nearby. The foragers travel away from the nest to search for seeds, and they may run into the foragers of a neighboring harvester ant colony. When ants of different colonies meet, they usually touch antennae and continue searching the ground for seeds.

To understand the dialogue between neighboring colonies we must consider the colony life cycle, the invisible underground part of the nest, and the daily round of work outside the nest. The internal organization of a colony determines how it grows and how it uses the space around the nest to collect food. This is the link between the network of interactions among ants inside their own nest, and the network of relations among neighboring colonies. Colonies send ants out to collect food, which is used to make more ants, who then go out and use more foraging space. The pattern of colonies in the landscape is generated by the work of ants deep underground.

2

THE GROWTH OF AN ANT SOCIETY

•

"You mean . . ." Horton gasped, "you have buildings there, too?"
"Oh, yes," piped the voice. "We most certainly do. . . .
"I know," called the voice, "I'm too small to be seen
But I'm Mayor of a town that is friendly and clean.
Our buildings, to you, would seem terribly small
But to us, who aren't big, they are wonderfully tall."

Dr. Seuss, *Horton Hears a Who*

•

THE COLONY LIFE CYCLE

The queen of an ant colony has a misleading name. There is nothing queenlike about her. Her job is to lay eggs, and she is important to the colony, as ovaries are important to a woman, but she has no special authority or privilege. There are no kings.

All ants begin life as eggs, then become tiny, wormlike larvae, and finally turn into pupae, resembling ants enclosed in a papery case. When an ant emerges from the pupal case it is an adult and does not grow any more. Upon emergence, the ant is one of three kinds: one of the ants you see walking around, which are sterile female workers; or one of the winged reproductives, males or queens. Only queens mate and only queens can produce females, either workers or queens. But workers, who do not mate, or queens, who do, can both produce males because males come from unfertilized eggs.

Red harvester ant queens generally live for fifteen to twenty years. Male ants live for only a few weeks. Worker ants live about a year in the laboratory, but in the field, we have seen live marked workers no more than six weeks from the day they were marked. However, the ants that I marked were already working at exterior tasks such as foraging, and if harvester ants are like other species, foragers are the oldest workers in the colony. Thus the six weeks that marked foragers lasted were probably the final stage of a much longer lifespan.

A new ant colony is formed when a male from a parent colony mates with a female, or queen, from another parent colony, and the newly mated queen then founds a new colony. Using only the sperm from that mating, that queen will produce all of the ants in the new colony. Her fertility is especially impressive because she must go on pumping out ants for 15 to 20 years to maintain the colony throughout her life. Once the queen dies, the colony will not adopt another queen or take back one of its own newly mated daughter queens. Instead, the surviving workers will eventually die and then, with no one to make more worker ants, the colony itself will be dead. Newly mated queens never go into existing colonies; their only option is to begin a new colony.

Males and virgin queens mate once a year, in the summer, after the rains. Existing colonies produce winged virgin queens and winged males, both called "alates" (from Latin *ala,* wing). All the colonies in an area send their alates to the same mating flight. A population of colonies comprises all the colonies whose reproductives are within mating range of one another.

Each summer, once the rains begin, the alates peek out of their nests every day at the end of the morning activity period—and often the workers pull them back in. Then somehow, one afternoon, they decide to go for it. It is usually the first clear day after the second or third heavy rain. Probably the ground must be softened by the rainwater before the queen can dig a nest. How do all the colonies in an area arrive at a consensus about when the mating flight will occur? The alates emit a strong sweet odor. Perhaps once a few females fly, the males detect the odor and decide to fly too, and as the cloud of odor passes over other nests more alates take off, creating an ever-thicker cloud. Suddenly the ground is covered with excited workers circling around their nest mounds, and the air is filled with the buzz of the large, slow insects taking off into the air.

The alates fly to some spot on the ground or onto low bushes. Where they go changes from year to year; perhaps large aggregations end up wherever the first few alates happened to land. In some years the mating aggregation is in several large patches, but in others thousands of alates all seem to have gathered in a single place, covering many

At a mating aggregation, a winged queen (in the middle) mates with one male (below) while another, holding on to her thorax (above), may be next to mate with her.

square meters with scrambling, copulating winged ants. The males rush onto the females. The females run along the ground, often each with one male attached to her abdomen and several more riding her thorax and head waiting for a turn; queens often mate with more than one male. Eventually a queen shakes off all the males and flies away.

By evening, males huddle together for warmth in clusters under a few bushes. In this species, as in many others, it is easy to tell the males from the queens because the males

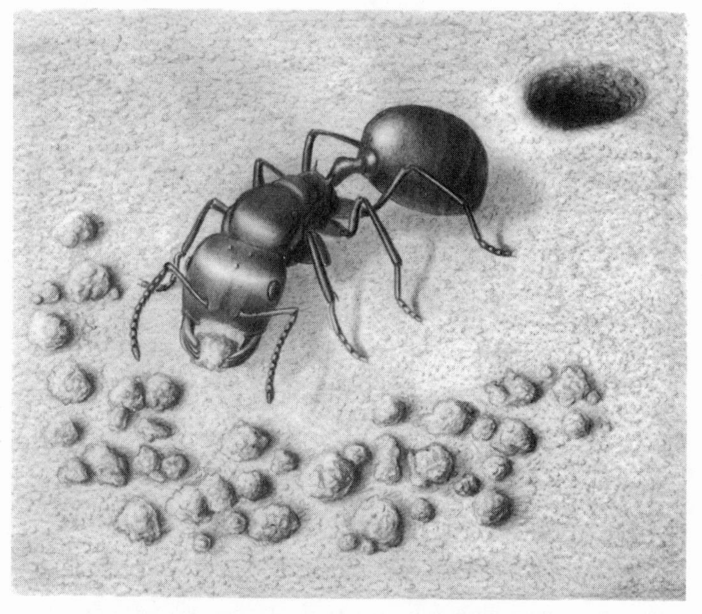

After mating, the queen drops her wings and begins to dig a nest.

have much smaller heads. They live for such a short time that they do not even need to eat, so their tiny heads lack the jaw muscles that females have. The next morning the males will be clustered under the bushes still, now seeking the shade, but within a day or two they all die.

A newly mated queen lands somewhere and runs around awhile. Most are eaten by birds and lizards, or dragged off by ants of the same species or by the large *Aphaenogaster* ants. But some manage to drop their wings and begin to dig a hole. Every few minutes the queen comes up with a clump of dirt in her mandibles, adds the clump to the crescent of dirt next to the hole, and disappears down the hole again. She digs a tunnel about 12 to 18 inches deep the first night, and she never emerges again. The morning after the mating flight there are still wingless queens around, but by the next day, those that did not manage to dig a nest have been eaten. Within a few days, the open nest-holes are plugged, the crescents of dirt have been swept away, and there is no sign that the mating flight ever happened.

In the lab, tiny workers emerge from the queen's first batch of eggs about six weeks later, and a few observations suggest it takes about as long in the field. The first workers are fed by the queen using her own fat reserves. They are tiny, but able to collect food and care for the second batch, who are much closer to normal worker size. These first workers emerge to forage in the fall, before the colony shuts down for the winter, between about November and March. But a new colony does not become a presence in the neighborhood until the following spring, its first, when it may have 500 workers in a tiny nest under a bush.

By the time a colony is two years old, it may have a thousand or more workers, an elaborate underground nest, and a small number of ants that forage outside the nest. Once a colony survives to be two years old with the queen intact, it is likely to live another 15 years. Though many thousands of queens mate on my study site each year, and many hundreds live long enough to dig a nest, only 20 to 50 survive to produce 1-year-old colonies the following summer. The mor-

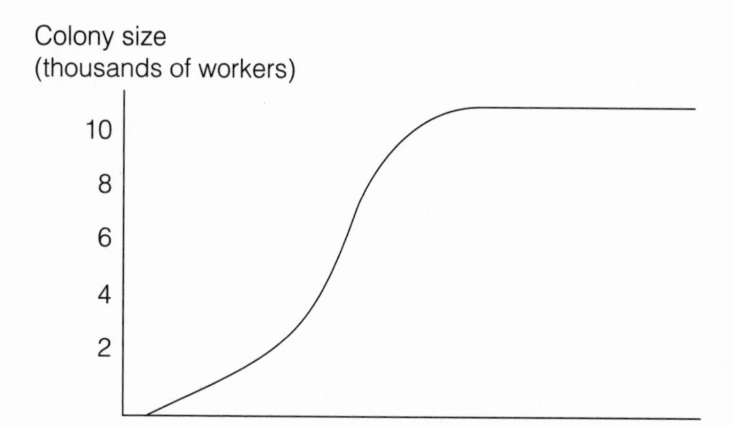

As the colony gets older, it grows from a single queen to thousands of workers. It reaches a size of about 10,000 workers when it is about five years old, and remains at this size for the rest of its life. When the queen dies, at the age of 15 or 20 years, the colony dies.

tality rate for alate queens is over 99 percent. The mortality rate for 2-year-old colonies, however, is very low, about 5 percent.

Colonies begin to reproduce, making the alate males and virgin queens to send to that year's mating flight, when they are about five years old. By that time they have reached a size of about 10,000 ants. A colony will stay at about that size for the next 10 or 15 years, until the queen dies (see chart, above). Some colonies continue to produce alates throughout their lives.

Much of a colony's success hinges on where a founding

queen digs her hole, though there is little evidence that queens choose carefully. The neighborhood of a new colony, especially the ages and sizes of its neighbors, determines whether the colony survives at all and if it does, how successfully it can reproduce.

THE CENSUS

I learned about the life cycle of a colony by watching colonies I know, year after year. Each year I take a census of the colonies on my site to find out which ones died and add

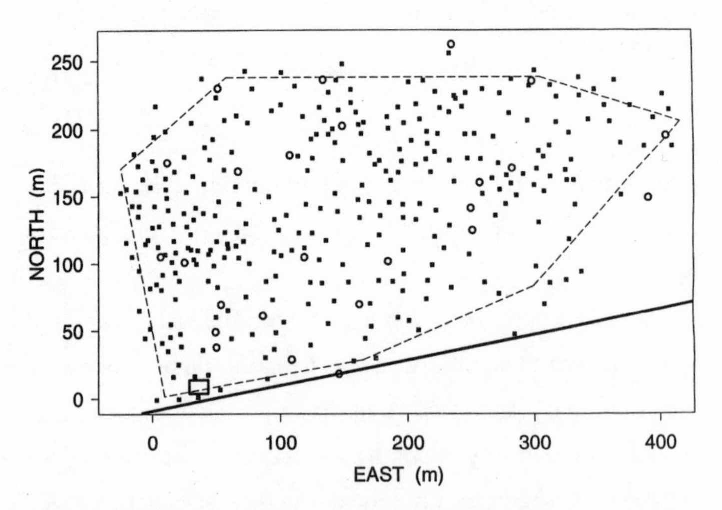

This map shows all colonies at the site in 1992. Each small filled square represents a colony 2 years or older; open circles represent 1-year-old colonies. The rectangle in the lower left shows the site of that year's mating flight. The diagonal line across the bottom is a road.

the new ones (see map, below). There are about 250 to 300 colonies on the site. Every one is marked with a numbered rock, and each year I find them all again, measure the location of each colony and put the new ones on the map. The census data tell us how old each colony is, and let us learn how colony behavior changes as the colony grows older.

At first I thought of the desert as a kind of repository from which I could select more or less identical colonies for experiments. But as I visited the same place year after year, I began to distinguish certain colonies and patches of the landscape. After about four years I realized that colonies change their behavior in interesting ways as they grow older and larger. That is when I decided to track colony age with maps that could be used to follow labeled colonies from one year to the next. We started out using a triangulation method, until I figured out that no amount of trigonometry would change the fact that if the location of one colony is measured as its distance from another, error will accumulate at an astonishingly high rate. We also wasted a lot of time bickering in the hot sun about sines and cosines. We gradually evolved a method of measuring locations from a series of permanent markers, using a tape measure and compass. This required people to add and subtract; if a colony was 32 m east of a 100-m marker, then its east coordinate was 132. Such calculations became difficult at the hot end of the morning, and some people seem to have a bias toward calling east west, or north south, leading to endless arguments later in the afternoon when we tried to put the newly measured colonies on the map and found them in strange places way off the site. Then new technology made it much cheaper and therefore possible for me to buy an infrared theodolite, which sends out an infrared beam to a reflector and calculates the location of the reflector from the time it takes the beam to get back. Now two of us can do in a few days the measurements that used to test the tempers of three or four people for several weeks.

I made the first accurate map in 1985, locating all the colonies in an area of about 8 hectares bounded on one side by the road, with arbitrary boundaries on the other three sides. (Metric units like hectares are the required standard for reports of scientific work.) I gradually expanded the study area each year until 1991, when I settled on the current region of about 12 hectares, about 300 m by 400 m. Since 1985, I have mapped all colonies on the site every year, about 300 a year.

The site has become consistently more crowded with *P. barbatus* colonies over the years. There are some especially crowded clusters of colonies of similar age. For example, the area we call "the inner city," in the middle of the west side of the site (see map on page 20) is crammed with colonies that first appeared in 1988, and have been jostling each other ever since.

Colonies are considered to be one year old the year they

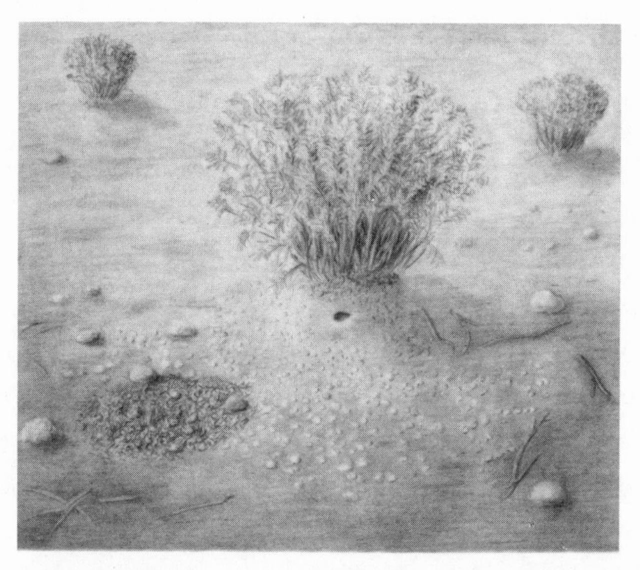

This is the nest mound of an older colony, 5 years old or more. The mound is about a meter wide.

This is the nest mound of a young, 1-year-old colony. The distance from the nest entrance to the edge of the midden (in the foreground) is about 10 cm.

first appear. The easiest case is a small colony in a location where there was no colony the preceding year. Such a colony was almost certainly founded by a newly mated queen after the previous year's mating flight. If there is a colony in the same place the following year, I add a year to its age.

What makes the census more complicated is that a colony sometimes moves out of its nest to another, newly constructed one. Ants, like people, are harder to count when they relocate, and about 10 percent of colonies move from one nest site to another each year. They move mostly after the summer rains, perhaps because this is when the soil is temporarily moistened, making it easier for them to dig new tunnels. Moving is a lengthy and dramatic process. For a week or two the ants are hard at work constructing one or more new nest entrances, each leading to a different, prospective new home. About half the colonies that start new nests then give up on the whole endeavor and stay where they were. Perhaps the new nest is judged to be no better than the old one. If the colony does move, a long, slow parade of workers makes its way to the new nest. Some carry other workers or immature ants, the larvae and pupae. After many hours, the queen will come out, just another ant in the stream. After all the ants have moved in, some ants go back to the old nest, carrying the seed stores to the new one. Moving the seeds can take many days. Sometimes another colony intervenes. The colony that owns the seeds may back off and begin foraging elsewhere from its new nest, while the intruding colony establishes a foraging trail to the plunder. On trails used to move stored seeds, traffic is always heavier and the ants more aggressive than on ordinary foraging trails; stored seeds after all are the end product of a great deal of sorting and processing.

Oddly, the same colonies tend to move year after year. This propensity makes it easier to keep track of nest relocation. Perhaps colonies move to get away from some fungal or bacterial infection, but then they take it with them and so have to move again. Relocation does not much change the distances between neighboring colonies, so colonies probably do not move to get away from one another.

INSIDE THE NEST

The effects of aging in an ant colony are probably the effects of colony growth. But how fast do colonies grow? The only way to find out is to count all the ants in a colony of known age. This means digging up a nest to find all the ants, which is no joke: Excavating a red harvester ant nest is a huge operation. The ground is like concrete for 2 or 3 meters down, and below that is a layer of large, dry rocks. We begin early in the morning, before the ants come out of the nest. We hire a backhoe to dig a trench about 2 meters deep along one side of the nest mound. Then, standing in the trench, wearing rubber gloves and with every crack in our clothing taped up to ward off the rain of stinging ants, we use trowels to lift off the top layer of dirt. As the ants are exposed, we put them into a container. The ants are poor climbers and can't get out of any container whose top curves in slightly.

The nest chambers are 3 to 5 cm across, with beautifully curved ceilings and walls, and flat bottoms, all lined with damp soil that dries to an adobelike surface. The chambers strongly resemble the nearby cliff dwellings built in the twelfth century on a larger scale, but with similar materials, by Pueblo people. Tunnels connnecting chambers can be many ant-widths across, but sometimes they are so narrow it looks as though even a single ant would have to squeeze to get through. The mass of chambers is roughly cone-shaped and about as deep as the mound is wide: maybe 25 cm for a 2-year-old colony, and 1.5 m for a 6-year-old one. Some of the chambers contain seeds stacked in neat piles, often the long, narrow grass seeds layered on the bottom and the round hard ones on top. Some chambers contain brood, with the larvae laid carefully on the floor of the chamber.

Once the exposed chambers are removed, we are standing in the trench facing a bare hole in the rock, with no sign yet of the queen or of most of the brood. All the ants have vanished. They have escaped, sometimes another 2 m down, by a narrow tunnel to a few chambers 2 or 3 m deep in the rock. The ants probably begin to move the brood, the newly emerged workers, and the queen, down to the bottom of the nest at the moment the first crunch of the backhoe shovel warns of impending catastrophe.

I learned to excavate harvester ant nests with the help of Stefan Cover, Curatorial Assistant for the ant collection at the Harvard Museum of Comparative Zoology. Stefan has accurately named the three stages of an excavation: Hope, which takes us through the first hour, while we lift off the soft dirt of the upper nest; Existential Despair, lasting four to five hours as we hack away with pickaxes and shovels in an apparently barren rocky hole; and then, finally, Anger, when after all of the sweat and ant stings we think we are damn well going to get the queen and make it worthwhile. When we excavate nests to count the ants, finding the queen is crucial, because only then can we be sure we have counted all of the ants that moved with her to the most remote chambers of the nest. When we are digging up a colony to bring it back to the laboratory for study, finding the queen is essential because a colony without one is short-lived; only the queen can produce more workers when the current ones die. Whoever finds the queen gets to name her and, by extension, the colony. The best way to guarantee success is to start early enough to get to the Anger stage before it is searingly hot, while there is still another energetic hour or two for chiseling away at the bottom of a 2-m-deep rocky hole in pursuit of any tiny crack that may be the escape tunnel. It also helps to have people along who will stand around saying, "The queen's in there somewhere, I know she is"—especially Stefan, whose hunches about which patch of rock the queen might be in are usually correct.

HOW MANY ANTS IN A COLONY?

To count the ants in colonies of known age, we had to destroy some of the colonies whose ages were known from the yearly census. Some of the excavated colonies survived in the lab, but so many ants died during counting that the depleted colony would be at a great disadvantage if we had put it back on the rubble of its excavated nest. This disadvantage would bias data on the survival of colonies in natural conditions, so no excavated colonies were put back.

In many species of ants, the number of workers grows very rapidly when the colony is young, and then levels off. In a red harvester ant colony, growth levels off when the colony is about five and begins to reproduce. We excavated colonies of known age ranging from two to four and counted all the ants, pupae, and larvae (see chart on page 19). Bill MacKay, now a professor at New Mexico State, dug up and counted 226 harvester ant colonies for his dissertation research. The colonies were of unknown age, and all appeared to be "large." The largest colony either of us ever found had about 12,000 ants, and most were closer to 10,000.

While it is possible that in the right conditions colonies could grow larger than 12,000, there is no evidence of it. Why don't colonies continue to grow indefinitely? It is amazing that a queen could continue to produce as many ants as she does, year after year, but perhaps there is a physiological limit on her egg-laying rate or sperm-storage capacity. This limit may set the maximum size of a colony.

Before excavating colonies, I measured their nest mounds to see if mound size corresponds to the numbers of ants in the colony. Nest mound size and the total number of ants active outside the nest do reflect fairly well the total colony size. Other measures, such as numbers of foragers (a subset of the ants outside the nest) do not correlate as well with true colony size. Though you can estimate the size of a colony from the size of its nest, the rule of thumb varies from year to year. In some years, most colonies will have large mounds; in other years, mounds tend to be small. Within a season, older colonies will have larger mounds (until mound size, like colony size, levels off at five years). But from year to year, no standard nest mound size can be used as a measure of colony age or size. Sadly, the only way really to find out a colony's size is to dig it up and thus destroy it.

THE UP AND COMING, AND THE IDLERS

A colony is divided into two broad groups: those that work inside the nest, tending the queen and brood, piling up stored food, and just milling around; and the exterior workers, who move between the world outside the nest and the chambers closest to the nest entrance, but rarely go down into the deeper parts of the nest. The exterior workers comprise only about 25 percent of the colony. At any time the interior and exterior workers are different ants, but it seems the interior workers will move up eventually to join the exterior workforce.

Several kinds of data support the idea that at any time, the interior and exterior workers are distinct groups of ants. MacKay marked exterior workers before he excavated nests, and rarely found them far below the surface. Anne Fullerton, doing undergraduate honors research in my lab, followed the movements of ants in laboratory colonies. Each colony was housed in a series of plastic boxes where the queen and brood usually stayed, connected to a single plastic box, the "outer chamber," that led to a large foraging arena. Ants moved between the boxes, and between the outer chamber and the foraging arena, but rarely between the boxes and the outer chamber. This suggests that one group of exterior workers went from the outer chamber back and forth to the arena, and that a second group moved around inside the nest. Few ants were needed to carry food from the outer chamber back into the nest.

In many ant species, ants work first inside the nest and later move to tasks outside the nest; this is also the usual trajectory for honeybees. The ants that work inside the nest may be younger ants, who first care for brood and store seeds, and only in the last weeks of their lives move up to tasks outside the nest.

The upper chambers of the nest, which feed into the nest entrance, hold the 25 percent of the colony that is working outside the nest. I have watched ants in these chambers using a fiber-optics microscope. The exterior ants are part of a relay that transfers food and waste between the deeper parts of the nest and the world outside. When foragers bring in food, they drop it in a chamber just inside the nest entrance. Ants come rushing up out of the nest, walking over seeds. Other ants come up from below, pick up the seeds, and carry them down to the ants that must make the neat piles of stored seeds we find in excavations. Some of the ants coming up carry dirt in their mandibles; these are the nest maintenance workers that come out, put the sand down near the nest entrance, and turn around and go back in.

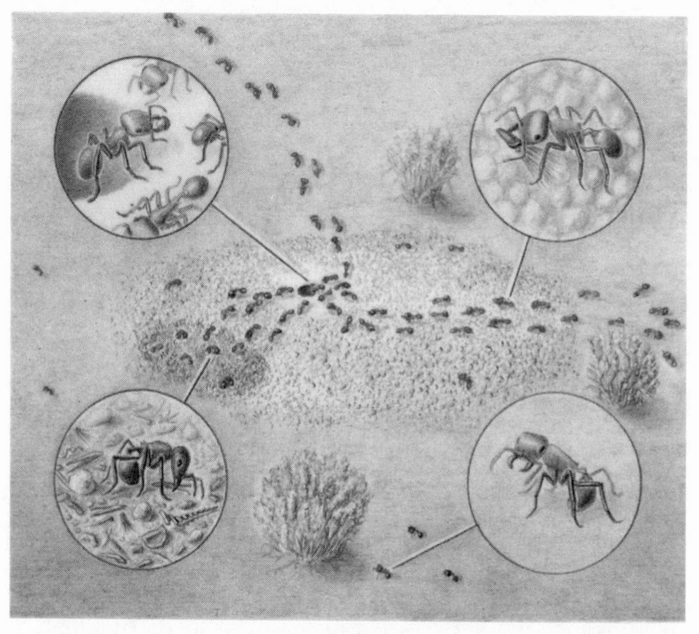

Four tasks are shown inside the circles; moving clockwise from the upper left, they are nest maintenance, foraging, midden work, and patrolling. The main scene shows ants of all 4 tasks active on and around the nest mound.

It is hard to believe that the thousands of ants inside are all hard at work. If not, many ants must be inactive inside the nest. There is a blurry distinction between reserves (ants that might come out if they were needed) and just plain inactive ones. A colony with 3,000 ants working outside doesn't need all 7,000 of the ants inside to feed the brood, store the seeds, and maintain the nest. Perhaps on the evolutionary timescale, events that suddenly require an extra 1,000 ants occur often enough to make it worthwhile to keep the reserves on hand. I have never seen such an event, but my seventeen summers of ant-watching are as nothing in the millions of years of ant evolution.

THE DAILY ROUND

A colony does its chores outside the nest in the same sequence every day. The daily round begins soon after sunrise (see chart on page 33).

The first groups of ants to emerge in the morning are the nest maintenance workers and the patrollers. Sometimes the nest maintenance workers actually open up the entrance, removing dirt that has washed in or that was used to plug up the nest entrance at the end of the previous day's work. In very dry weather, the nest may be open all night. The nest maintenance workers come out in small bursts of eight or 10. Each one carries a bit of soil in its mandibles and places it on the mound a few inches away, then returns immediately into the nest for another bit of soil.

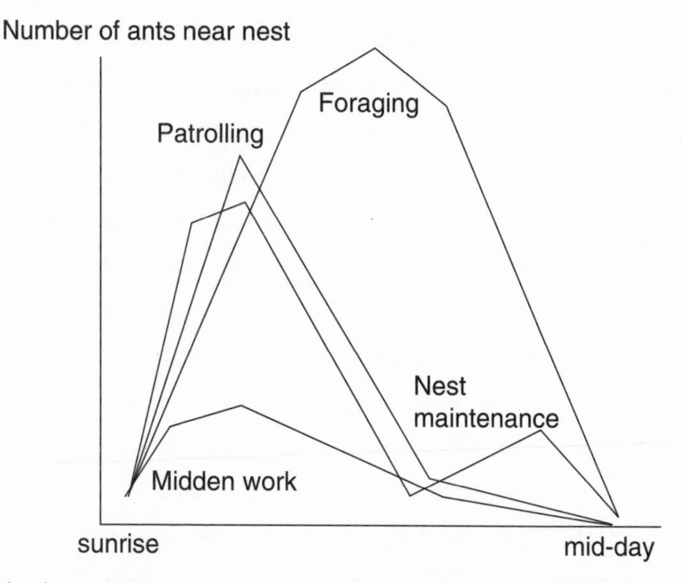

This chart shows the numbers of ants engaged in each task outside the nest, during the morning activity period. Nest maintenance workers and patrollers come out early in the morning, and later the foragers emerge. The chart shows only the numbers of workers near the nest, not the large number of foragers that travel far from the nest (see page 37). Midden work continues through the first half of the morning. As the last foragers go in, there is a final burst of nest maintenance work.

The patrollers do not look as determined as the nest maintenance workers. The first patrollers peek out of the nest entrance, holding on to the edge with their front legs and waving their antennae around to check out the chemical landscape. When they work outside the nest, patrollers have a characteristic way of walking, with the abdomen tucked under the thorax. When ants of different colonies meet, each ant inspects the top front of the other's abdomen. Perhaps a gland that exits on the top front of the abdomen secretes a chemical that patrollers use, and tucking the abdomen under is how the patrollers send out this chemical. Patrollers tend to walk in a zigzag trajectory rather than in a straight line. They search and interact more often than other ants, stopping frequently to inspect the ground or other ants with their antennae. At first they move around the surface of the nest, and eventually they spread farther away from it. Somehow the patrollers choose the directions to be taken that day by the foragers. There may also be some diplomatic transactions with the patrollers of neighboring colonies, to decide which colony will use a certain patch of foraging area.

The foragers begin to emerge as patrolling reaches its peak, when the patrollers are dispersed around the foraging area. Foragers walk out of the nest looking as if they know where they are going, and usually head straight off the mound and out into the surrounding vegetation. Discrete trails are more common some years than others, and some colonies tend to make linear trails while others fan out in wide sections around the nest. There is also a daily pattern of trail formation. Often the first few hundred foragers will fan out, but later on the base of the fan, at the nest side, becomes more linear, and eventually the foraging "trail" is shaped like a long blob with a narrow tail (see map on page 43).

The foraging period may last for several hours. By the time foraging reaches its daily peak, the nest maintenance workers and patrollers have gone back inside the nest. Another group, the midden workers, comes out with the patrollers and may go back in with them or stay active throughout the morning. Midden workers sort and pile the refuse pile, or midden. At times the middens are full of the husks of that week's most popular seed; at other times the ants sort through indistinct bits of dirt.

Midden workers seem to have a mission, but it is not clear what it is. Over the course of a week they may move an entire midden from one side of the mound to the other, and then move it all back again. I suspect that midden material acts as a carrier of some scent-marking chemical. When I removed middens from colonies of the harvester ant, *P. badius,* I found that ants of other species were more likely to intrude. *P. badius,* which lives in the pine barrens of the coastal southeastern United States, collects bits of charcoal, in the soil from frequent forest fires, to pile on its mounds. The harvester ants of the Southwest collect small pieces of fossilized bone, as well as pebbles (a habit that makes the ants popular with paleontologists). Both charcoal and bone are porous, inert substances that would be ideal containers for a scent-laden chemical.

The peak of foraging is when the air begins to get hot, so hot that you can hear the silence it generates. The foragers rush along, traveling sometimes 10 or 20 m away from the nest. When they reach the searching region of the trail they slow down and slowly antennate the ground. As soon as an ant finds food, it picks it up and goes straight back into the nest; very few (less than 10 percent) return without a food item. A forager's trip takes about 20 minutes, though we have seen unlucky foragers spend up to 138 minutes outside the nest.

Eventually the outgoing stream of foragers grows thin, and soon all of the foragers are headed back to the nest. There may be a final burst of nest maintenance; ants race out, put down their burdens and dash back inside the nest. Soon the glare of the midday sun shows a landscape scattered with silent, empty mounds. On the warmest days of summer the ants inside wait out the midday heat and come out again in the evening for another round of foraging.

I learned about the daily round of the colony by making hourly observations of many colonies, going around the same colonies hour after hour to count the numbers engaged in nest maintenance, patrolling, foraging, or midden work. The drawing on page 37 was made by tracing the paths of ants from a film. We made this film in the mid-eighties, when a video camera was beyond the resources of a graduate student, using an old movie camera that had to run off a generator. Kristin Roth and I lugged the generator on a pallet for what seemed like miles across the study site. While trudging along, we found ourselves suddenly looking down through the handles at a large, coiled mojave rattlesnake on the ground. We just kept walking; the generator was so heavy that it did not occur to us to put it down, and have to pick it up again, for a mere snake. Then we carried out a (much lighter) chair, and to get the whole mound into view I stood on the chair to point the camera at the mound.

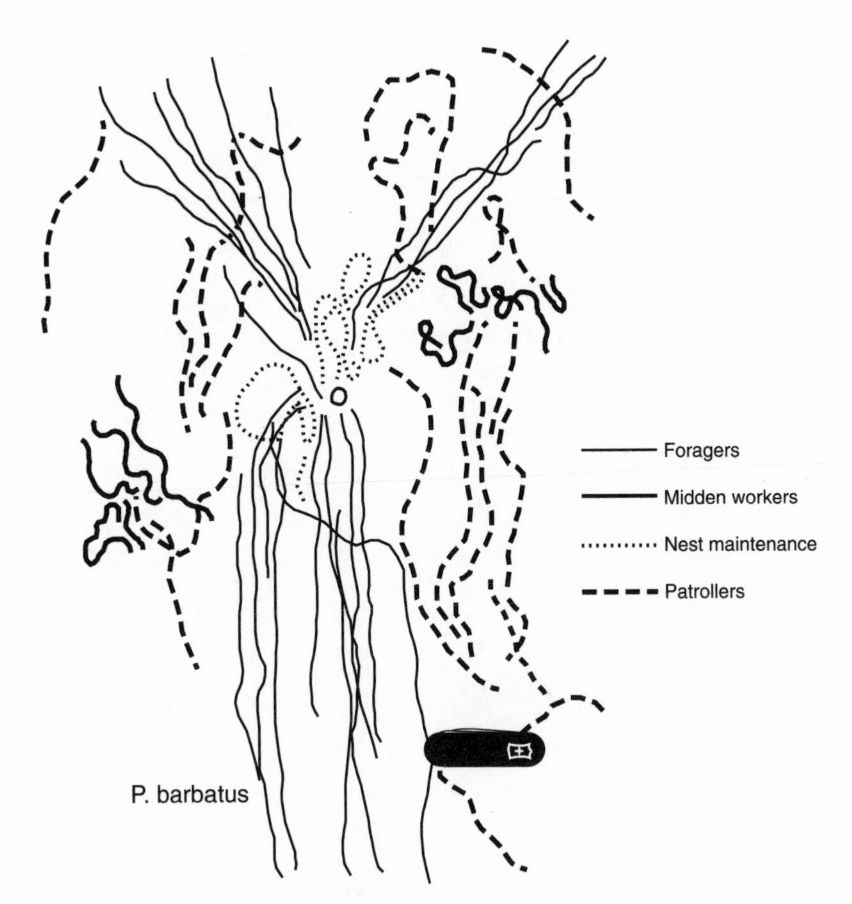

The paths of ants performing each task were traced from film. The open circle in the center is the nest entrance. The thin solid lines show the paths of foragers on three trails. The dotted lines show the paths of nest maintenance workers as they come out of the nest, put down their load of sand, and go back in. The thick solid lines are midden workers on two separate middens. The dashed lines are patrollers, at this time of the morning going around the outer edge of the mound. In the lower right is a Swiss Army knife, for scale. Each task is performed in a different place, but ants of different tasks meet as they go in and out of the nest entrance.

As the drawing on page 37 shows, the tasks are performed in different places, and that makes counting easier. I use a hand counter and count one group at a time. For the nest maintenance workers, I count all the ants that leave the nest and keep counting until the last one gets back into the nest. They tend to come out in bursts, and while there may be more than one group at work, this is a way to limit the times I count the same ant. For the patrollers, I walk around the nest, counting every patroller I see within a radius of 1.3 m from the nest entrance, and I stop when I get back to the place I started. For the midden workers, I count all the ants currently moving objects around on the midden or standing on the midden wiping down their antennae with the comb on their forelegs. For the foragers, I count all of the ants on any foraging trail, or in any fan, within the same radius. This spatial limit, just outside the edge of the largest mounds, means the count vastly underestimates the total numbers of active foragers (so I measure that total in other ways), but it provides a measure of daily rhythms in numbers foraging.

I have tried to assess the accuracy of these counts in two ways. The first is counting the same colony over and over, at more or less the same time, and comparing the counts. The second is to have several people make counts at the same time. Both these measures provide counts that agree within 5 percent. In 1990 I estimated that I had at that time made such counts 10,000 times. Whatever the inaccuracies of the method, they are fairly constant, so that such counts can be used to compare colonies.

The daily pattern I have described is a typical one for a hot day in the summer. Colonies differ, and changes in the weather from one day to the next cause the pattern to change. In fact, day-to-day variation exceeds colony differences: On some days, most colonies forage in great numbers and for a long time. On other days, such as the hot days before the summer rains, foraging is sparse.

The beginning of the day's surface activity seems to be triggered by light and temperature. Colonies with entrances that are first to catch the morning sunlight emerge a little sooner than those with entrances covered by early-morning shadow. Once a colony is active, there is a domino effect; the work of one group seems to elicit the activity of the next.

Each colony carries out its daily round in a neighborhood of colonies, and neighborhoods differ. Groups of grand old matronly colonies, spaced far apart, do not deign to notice each other; a cluster of tiny new colonies seem unaware that each has foragers on another's doorstep; colonies in a neighborhood of 3- or 4-year-olds keep pushing and shoving at one another's foragers. Or this is how it looks. Let's consider in more detail neighbor relations and why they matter.

3

FOOD AND THE FOREIGN RELATIONS OF ANT SOCIETIES

•

Look to the ant, thou sluggard;
Consider her ways and be wise:
Which having no chief, overseer, or ruler,
Provides her meat in the summer,
And gathers her food in the harvest.

Proverbs 6:6

•

WHICH COLONY IS THE BEST FORAGER?

The top few centimeters of soil hold the seeds that harvester ants must find, get back to the nest, and store. What looks to me like a bare patch of desert is the source of an ant colony's livelihood. A colony shares the desert with its neighbors, and one that extracts the seeds from the soil more effectively than its neighbors might have a crucial competitive edge. Foraging activity seems to fluctuate from day to day, and from colony to colony. Do harvester ant colonies compete for food, and are some colonies better foragers than others? The first step in answering these questions was to find out how much of the fluctuation in foraging behavior is due to the unusual abilities of certain colonies.

On some days the entire desert seems filled with foraging harvester ants, and on others, almost everybody stays indoors. We counted the foragers of thirty-seven colonies, between 8:00 and 8:30 in the morning, and found strong day-to-day variation in the numbers of ants out searching for food. Colonies send out different numbers of foragers, but the most striking trend is that all the colonies tend to vary together. These waves of foraging intensity correspond to the weather. On the day after a storm, when flooding has damaged many nests, few ants forage and many do nest maintenance work. Once the nests are repaired, foragers pour out enthusiastically, collecting seeds that were buried in the soil until they were uncovered by the floods that sweep the desert surface. Then, after some climatic event has inspired a lot of foraging, food stores may be so high that, in many colonies, foragers have little incentive to go out and forage.

However, some colonies forage more intensely than others. A crude measure of foraging intensity is whether the colony forages at all; no colony forages actively every day. So, every day for two summers, we noted whether each of about one hundred colonies was active or not. I found a colony-specific tendency to forage actively; the number of days a colony foraged actively one season was positively correlated with the number of foraging days the next year.

From day to day, colonies change *where* they forage, as well as how much. Are some colonies claiming all the best places to forage? I discovered that foraging sites change by mapping the foraging trails of the same colonies, day after day. The foraging map is a picture of foraging by one colony on one day. The ants may use linear trails that fan out at the end where the ants search for trails, or the fan may be a wide blob with its base at the nest (see illustration below).

To make foraging maps, we put out grids on the ground to use as points of reference. At first we made star-shaped grids out of string, attached to the ground with tent pegs, centered on the nest mound. They looked like huge spider

This map shows the foraging trails of a colony on one day. The nest entrance is in the center.

webs spread across the desert. We spent a long time winding and unwinding the grids each morning (we could not leave them up overnight because a cow might come to grief in one), and even when they were wound up and stored in a box, the arms of the web kept trying to entangle themselves. So the star-shaped grids evolved into rectangular arrays of small plastic flags placed at 6-m intervals, laid out with a compass, a tape measure, and much bickering about the fundamentals of geometry. It seems as though there should be a clever way of getting long rows of evenly spaced flags by filling a small number of triangles, without having to do so much measuring—but there isn't. Some summers the study site has been covered with patches of orange, yellow, blue, and white rectangular flags, the colors cows seem least prone to eat, making the site look like a rugged, cowboy-style miniature golf course.

We developed a set of mapmaking criteria, which we use to mark the location of foragers on a piece of graph paper that has the grid points marked to scale. A place is recorded on the map as a foraging trail or foraging fan when there are at least five ants in 20 seconds crossing an imaginary 5-cm line at that place. A place is recorded as the edge of a foraging trail or fan or blob if there are no ants within 1 m in 10 seconds. A colony's foraging trails tend to extend farther from the nest, and cover more area, as the day goes on, until foraging activity peaks in late morning. Each colony mapped is visited at least twice, and the largest area mapped is the one we retain. Many students have learned to make foraging maps over the past eight years. A new person becomes a qualified maker of foraging maps when he or she can produce a map of the foraging area of a colony that looks like someone else's map of the same colony, without any consultation; this usually takes two or three days of practice once a person is used to looking at harvester ants.

The foraging maps of the same colonies, day after day, made it obvious that colonies change where they forage from one day to the next. Around their nests, harvester ant colonies sometimes clear pathways through the grass to be used by foragers. The cleared pathways are sometimes called "trunk trails," a term unfortunately used to describe very different things for different species of ants. Cleared pathways are more common the summer after a year of heavy rainfall, when there is more vegetation. Some colonies always have at least one or two cleared paths, and other colonies never do, regardless of the amount of vegetation around the nest. A neighborhood of colonies in a time-lapse movie filmed from above might look like a group of starfish stuck on a rock, or a group of amoebae. Each colony's foraging trails wave around, extending and retracting like pseudopods, and occasionally the arm of some colony meets that of a neighbor. An established colony seems to have up to about eight possible foraging directions. On any day, however, it uses only three to five of them. A colony's pathways, cleared or not, are not all used every day.

Why do colonies forage in different places from one day to the next? An obvious reason is that they are tracking the food supply. Avoiding neighbors might be another reason to shift foragers around. Let us consider these possibilities in detail.

DO COLONIES KNOW WHERE TO SEARCH?

The species of ants that tend to invade people's houses beautifully illustrate a rapid response to a change in the natural resources of the landscape. A tiny crumb of cake lands on the kitchen counter and the ants seem to be there within minutes. Indeed the observation that where there is a picnic, there will be ants, rests on the notion that there is an ant lurking everywhere, all the time, ready to mobilize its nestmates when a picnic appears.

Harvester ants, by contrast, are slow ants, foraging in an environment that changes slowly. Crumbs come and go on your kitchen counter, but a seed may be stuck at the soil surface for weeks or months. Seeds are durable, and harvester ants may keep them for years; certainly I have found, inside excavated nests, seeds that were produced at least eleven months before. It often seems that harvester ants have to be coaxed into taking an interest in food; it is much easier to persuade them not to forage at all than to encourage them to collect food bait, whereas the species of ants that invade my kitchen ignore all my efforts to dissuade them. Harvester ants need a reason to come out and forage, because foraging has a heavy cost for them. They obtain water from metabolizing the fats in the seeds they eat, but to get the seeds, and thus water, they have to use a lot of water traveling through a hot, dry desert. On many days, especially for a colony that has a lot of stored food already, it must be more worthwhile to wait in the cool nest, conserving water, than to go out searching for more food and the water that comes with it.

Harvester ant colonies gather food rather decorously, each phase at its proper time, without any appearance of undue eagerness. They will go back to the nest and quickly recruit others to gather food, as more enterprising ants do, but only at certain times and places. If I put out an abundant source of food, such as a pile of birdseed, the ants may form a foraging trail to retrieve it. But foragers will come to the food only if the patrollers encounter it first, during their early morning search period. If the patrollers find the bait at the right time, hundreds of foragers appear and a large pile of seeds will be whisked back into the nest in a few minutes. The same birdseed put out later in the day, after the patrollers have gone back in and the foragers have settled on the day's foraging directions, will be ignored. Foragers will walk right over the birdseed on the way to one of that day's foraging areas, where the same ants may have to search for twenty minutes to find a tiny buried husk with nowhere near the nutritional value of the giant oat flakes or millet seeds I offered.

Piles of mixed birdseed don't appear often in the foraging

area of a colony. But each day, a colony could adjust its foraging area to include regions that seem sufficiently rich to warrant a day's worth of searching and collecting. Patrollers apparently make this decision. They begin each morning, before foragers become active, by searching the nest mound and then venturing farther out into the colony's foraging area. By marking nest and trail patrollers, I found that the same individuals tend to patrol both areas, which means the same patrollers probably move from nest mound to foraging area each morning.

On one day we mapped the locations of patrollers, and then the locations of foragers. Then the next day we repeated the process. Comparing the distributions of forager and patroller directions, we learned that foragers use the trails explored earlier the same morning by patrollers. In this sense, patrollers choose the day's trails.

But how do patrollers tell foragers where to go? One might assume that explicit directions are necessary. This is not certain. The probability that a patroller stays in an area, rather than moving quickly away, might increase according to the rate at which it finds food. Even if patrollers are only slightly more likely to stay in the same place in response to slightly more food, the net effect would be for patrollers to gather in areas where there is more food. Then the first foragers might be slightly more likely to search areas where they encounter more patrollers. Finally, subsequent foragers simply have to follow the paths taken by the foragers that just entered the nest.

Patrollers may not choose the best possible sites, but rather the trails that are just good enough. Particular trails might not be consistently bountiful sources of seeds. A mature colony has up to eight usual foraging directions, established by repetition. Patrollers might explore the colony's habitual trails at random, and if they find a sufficient quantity of seeds somewhere, they remain there longer. If the patrollers stay longer where they find food, and foragers go only to the trails that contain a threshold number of patrollers, then foraging trails would track at least some minimum level of food abundance.

Seeds are a long-lasting food source, often buried in the top layer of soil. A region rich in seeds today should probably be rich in seeds tomorrow, unless a competitor with superior skills in tracking food is more likely to exploit a rich patch than a poor one. If food persists from day to day, then colonies could remember food abundance from day to day, somehow using today the trails that were most productive yesterday. For two summers I tried without success to induce colonies to choose the best trails of the day before. For many days running I put out abundant seeds at a place where a colony did not have a habitual trail. Ants recruited others and a foraging trail formed to collect the seeds. When I stopped putting out the seeds, the ants stopped foraging in the direction of my seed bait. Younger colonies were more likely than older ones to keep sending foragers in the direction where the seed bait had been, but no colonies of any age preferred my trail, an abundant food source, to their own. Patrollers do not seem to send foragers to the best source of the previous day.

It is possible, of course, that the trails I think the ants should prefer, because of the rich food I offer, are actually not superior to the ones the ants choose on their own. This example illustrates a basic problem of optimality models. Many biologists have told us what foragers ought to do, how they should maximize energy gained, or minimize time spent working to gain energy. But harvester ants have consistently failed to meet the predictions of optimal foraging theory. Is this because they are not optimal foragers, or because our models are wrong? They could be foraging optimally after all, for instance ignoring my birdseed to obtain some obscure mineral found only in dry husks of a rare grass. Our models, based on measures of caloric value of seeds, energy required to carry a seed of a certain size, and so on, miss this possibility completely. Or the ants may not be foraging optimally; they may be wasting time and energy to get low-quality seeds. We can't tell if some crucial piece is missing from our view of what is best for them.

Let's suppose that colonies just make do with foraging in any direction that has enough food to attract some threshold number of patrollers. This may be all the tracking of changing food abundance that is required in an environment where the distribution of food is random and doesn't change quickly. In that case, it doesn't matter much where the colony goes anyway. By contrast, if really good patches of food tend to pop up and disappear quickly, a colony needs to be able to find those good spots accurately and before the competition gets there. Even though I couldn't make colonies track the food sources I offered, perhaps they are following a changing distribution of food sources that I don't see.

I set out to learn how day-to-day changes in the location and intensity of foraging reflect the way food sources are distributed around the nest. The plants that produce the seeds that harvester ants eat are not scattered evenly. Some plants might be more nutritious, or some areas might contain denser vegetation and thus more abundant food.

I made a vegetation map of the study site, by identifying all the different plants I could find, and mapping the site according to the plants present in high densities. Small shrubs, such as Mormon tea (*Ephedra*), snakeweed (*Gutierrezia*), and *Isocoma tenuisecta,* grow throughout the study site. I divided the other vegetation into four types. The first type consists mostly of small herbaceous flowering plants (such as *Baileya multiradiata, Talinum aurantiacum, Solanum elaeagnifolium,* and *Allionia incarnata*). The second type consists of a variety of short grasses, many of the genera *Aristida* or *Bouteloua.* The third type consists of one species of very thick grass, *Hilaria mutica;* here the small shrubs do not grow. The fourth vegetation type is bare ground.

In the first summer of this study, we looked at the match between what plants were growing in a region and the seeds that foragers collected there. To find out what seeds are available to ants in each vegetation type, we needed help from the ants. I can't see the seeds that ants collect just by looking at the ground, though the seed is usually visible once it is in an ant's mandibles, especially if it's a seed of a contrasting color. I could sieve the soil for seeds, but other work shows that soil samples vary so much, from one square centimeter to the next, that no scheme apart from sampling everywhere would give credible results. So I used their ability to find seeds to let the ants tell me what species was found where. I identified the seeds collected by ants foraging in a particular place. This method doesn't allow me to distinguish the effects of availability and preference; it is likely that ants select some kinds of seeds from a larger pool of available seeds. But the ants' selection shows the match between what they collect and what plants are growing nearby.

We collected seeds from returning foragers in each vegetation type, using an aspirator, a little collecting device that entomologists of all sorts use. The device is a small plastic vial with two tubes leading into its stopper. Suck air from one tube, and a vacuum is created in the vial that can suck an insect into the vial through the other tube. Lurking next to the foraging trail, we collected ants we could see returning with something in their mandibles. The ant falls into the bottom of the vial clutching its treasure, but if you tap the ant's head gently with a twig it usually opens its mandibles out of sheer astonishment, dropping the seed. (The exception is when the ant has had the amazing good fortune to find a termite; then nothing will induce the ant to let it go.) Seedless foragers were released, to begin their search again or return, puzzled, to the nest. We collected seeds from 100 foragers from each of 24 colonies, in four weekly bouts in July and August in the tenth of my seventeen summers at the site. I sorted the seeds into types, giving them names of my own, like "chili pepper," "cantaloupe," and "conch shell," and then counted how many of each type had been collected by each colony. Tom van Devender, from the Sonoran Desert Museum in Tucson, identified all the seeds.

For two summers, I spent every afternoon staring at seeds under the microscope. I was impressed by both the beauty and the variety of seeds; by the ants' ability to find, presumably by smell, seeds buried in layers of mud; and by the preponderance of apparently worthless crud selected by foragers. Most of the seeds were only a few millimeters wide, but there were seeds that looked like seashells, antique flying contraptions, and Platonic solids, as well as seeds that looked like more familiar large fruits and seeds such as melons, Brazil nuts, and ears of corn. A favorite seed of the ants, *Plantago patagonica,* is a nearly transparent ellipsoid almost invariably surrounded by many times its volume of dry dirt. A huge proportion of the items collected by ants were husks and other papery, inedible wrappings of grass seeds. Many of the husks carried out by nest maintenance workers at the end of the foraging period are probably such mistakes.

I found no relation between the seeds collected by ants and the kind of vegetation growing in the foraging area. It might seem obvious that ants must collect what is growing nearby, but (at least in the summer) they don't. I found 26 plant species on the site in July and August, and seeds from 36 plant species were collected by the ants, but there were only eight in common: Only eight of the ants' choices grew on the site. Regardless of vegetation type, all colonies collected mostly one species of seed, the grass *Bouteloua aristidoides.* This is a grass that sets seed in September and October, and it was never seen growing on the study site. Our collections were made in July and August. During the summer, ants are collecting mostly seeds produced many months before. The seeds are distributed onto the site from elsewhere, moved by the wind and by flooding.

The following summer we compared the seeds collected on distinct foraging trails of the same colony. Colonies differed from their neighbors in the kinds of seeds they collected, but any given one of a colony's trails yielded much the same kinds of seeds as any other trail.

So, the food a colony gets from any one foraging trail closely resembles the food it gets from another. The decisions that lead a colony to forage in one place today and some other tomorrow are not complicated tactical maneuvers. Theory tells us that colonies should recruit each morning to patches that represent a delicate balance of costs and benefits, but this theory simply does not apply to harvester ants. In fact, it doesn't seem to matter much where a colony forages, and wherever they go, a substantial proportion of foragers will bring back inedible bits that will soon be thrown away.

How foraging trails change from day to day probably has some relation to the food that patrollers encounter in their early morning searches. But if food supply has a strong effect on where a colony forages, it is difficult to tell where this effect originates, because it looks as though more or less the same food is available throughout a colony's foraging area.

The influence of food on foraging remains a mystery. It is clear, though, that relations with neighbors strongly affect the location of foraging.

NEIGHBORS WITHOUT FENCES

The harvester ant colonies on my site are fairly crowded. The average distance between colonies is about 9 meters. The underground nest of a colony is only about 1 to 2 m wide, and neighboring colonies do not meet underground. Above the ground a colony's foragers can travel up to 30 m from the nest. The colony's patrollers and foragers act as arms that reach out into the area around the nest to search for food and bring the food back to the nest. The patrollers and foragers of a colony may encounter the patrollers and foragers of neighboring ones.

Neighboring colonies are not likely to be genetically

related. Colonies do not bud off and establish daughter colonies nearby. Instead, in the annual mating flight the winged reproductives, males and females, from many colonies in a population, all fly to the same place. There they mate, and there the males die. The newly mated females, the queens, fly off apparently at random and search for a place to build a new nest. There is no evidence that a newly mated queen flies back to the area near the nest from which she came. So a newly founded colony is not likely to be closely related to the established colonies nearby.

How do a colony's neighbors influence where it forages? The answer is, most simply, by frequent meetings. From foraging maps I knew something of the pattern in day-to-day movements of a colony's foraging trails. Of course, colonies never exist on their own, so the baseline pattern is not established independently of the effects of neighbors. But since both foraging and encounters between colonies have visible locations, I decided to find out how foraging is affected by encounters in specific places. I pursued this question in two ways. One was to describe in images and numbers the slow, silent dialogues between colonies. The other was to interfere with the dialogues and see what changes ensued. Both methods led to the same conclusion: Old, large colonies and young, small ones have very different relations with their neighbors.

First we mapped the foraging trails of some pairs of neighboring colonies that seemed close enough to meet. We

The foraging trails of two neigboring colonies overlap in the shaded area.

made maps of the same fourteen pairs every few days for several weeks. For each pair of colonies I arranged the maps in chronological order, like a cartoon, to see if there was a story. Most happened to bump into each other occasionally, but I could see no pattern. For a few pairs of older colonies, though, there were incidents that looked like avoidance. The trails of the two colonies might meet, two slow trickles of ants that flow into each other so that the searching ants of the two colonies mingle (see illustration on page 57). Sometimes ants from different colonies meet, and usually they touch antennae, or one uses its antennae to touch the top of the abdomen of the other, and they both rear back slightly, as if surprised, and then continue to search the ground for seeds. After the trails of two older colonies met, both colonies foraged away from each other, and the space between them had become empty of foragers. But one group of three younger colonies did the opposite. Week after week, their foraging trails continued to meet, and often the ants fought.

When harvester ants fight, they grab onto each other with their mandibles and hold on. Often each ant clamps the other's petiole, the segment that attaches the abdomen to the thorax. They can tumble around clenched together for

When ants begin to fight, each grabs the other.

hours. Sometimes one ant succeeds in breaking the other into two pieces. Sometimes an ant dies while clamped on to another, but the mandibular muscles of a dead ant maintain their grip though the rest of the ant may break off. In a colony that has been fighting recently, it is not unusual to see an ant walking around with just the head of its attacker still attached to its petiole.

Next we staged interactions between colonies. We put down lines of seeds between pairs of older colonies. I chose pairs of colonies of about the same age, but of which one colony had more foragers and longer foraging trails than the other. The line of seeds began near the trail of the weaker colony, the one with fewer foragers and a smaller foraging area, and led toward a trail of the stronger colony. I did it this

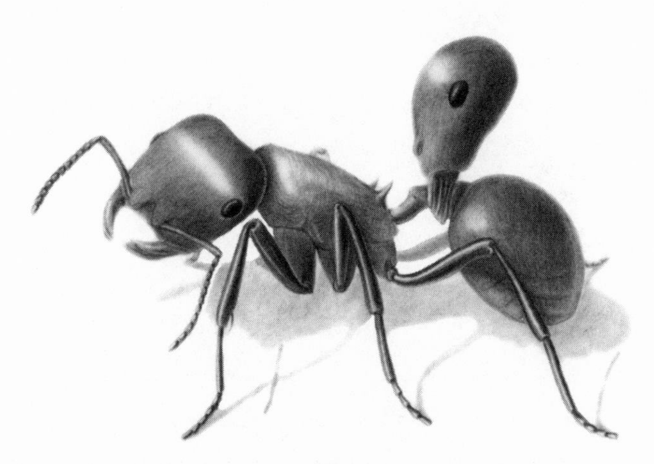

Sometimes one ant dies with its mandibles clamped onto another. The rest of the attacker's dead body may break off, but its head often stays attached.

way for two reasons: First, if the food were closer to the stronger colony, the weaker colony might never be able to get to it; and second, if the proximity of food were to bring out more foragers, I thought this increase would be easier to measure if there were fewer foragers to begin with. We put out seeds for ten days. In every case, both colonies foraged for the seeds. They often fought. Proximity to food did not have much effect on forager numbers: Though the weaker colony was closer to the seed bait, it still sent out fewer foragers. This experiment showed that colonies do not control exclusive foraging areas. Instead a colony will collect a rich food source even when it is closer to a neighbor's usual foraging area.

I repeated these experiments with pairs of two age classes. We chose pairs of older colonies, both more than five years old, and pairs of younger ones, both two years old. Here we put down a line of seeds equidistant between the colonies. The seeds were put down only on one day. We continued to map the trails for ten days after we had stopped putting out bait.

Younger colonies were more willing than older ones to put up with neighbors in order to get food. In all colony pairs, regardless of colony age, both colonies foraged toward the seed bait when I put out seeds. In both age classes, the colonies foraged toward the bait even after it was gone. But younger colonies kept on longer. Younger colonies continued to forage toward the bait and to fight with each other for up to six days after the bait was gone. The older pairs gave up the conflict sooner. Once the food was gone, they were more likely to direct their foraging efforts elsewhere. For the younger colonies, a site that had offered abundant food a few days before was still worth fighting over.

Weather, or something else that changes from day to day, seems to affect whether ants fight. On average, about 40 percent of encounters between colonies involved some fighting, but this average is over a range from 10 percent to 65 percent. On some days, fighting is widespread; on others, many colonies meet but fighting is rare. The fighting season seems to last for several weeks after the rains begin, when foraging activity is very intense. Perhaps flooding washes away from the ground some scent marks that normally inhibit fighting. Or perhaps the rainy season happens to create wind or humidity conditions ideal for some fight-inducing pheromone.

WHEN A NEIGHBOR VANISHES

To find out if older colonies are really as polite to each other as they often seemed, I did another experiment. I prevented one colony from foraging, and mapped the foraging areas of its neighbors. If interactions between colonies cause them to avoid each other, then when one colony disappears, its neighbors should shift their foraging trails. I then allowed the imprisoned colony to forage again, to find out whether it could regain its status in the neighborhood after an absence.

To enclose a colony we put a piece of aluminum siding around its nest mound, digging it in a few inches and leaving a circular corral about 2 feet high. Harvester ants are inept climbers, and they couldn't scale the fence. We fed the colonies every few days, giving them enough seeds to prevent effects of food deprivation, in case the enclosed colony's food stores were running low. Colonies that are fed a lot of birdseed day after day will eventually become satiated and stop foraging, so I didn't feed the enclosed ones much. The enclosure, then, prevented a colony from foraging, but otherwise didn't harm it. The ants in enclosed colonies worked at digging tunnels under the fence, but we buried the bottom edge of the fence again every day before any foragers could escape under it. For their neighbors, the enclosed colonies simply vanished.

The first time I did this experiment, I used five sets of neighboring colonies. Each set included one enclosed colony and three or four neighbors, for a total of 18 colonies. All the colonies were at least 5 years old, except for three unusually large 4-year-old colonies. On the day before any colonies were enclosed, day 0, we mapped the foraging trails of all colonies. Then in each set, one colony was enclosed for 15 days. We mapped the foraging trails of the neighbors on days 5, 10, and 15. On day 16 the enclosed colonies were released, and we mapped the trails of all colonies again.

The neighbors shifted their trails into the foraging areas

of the enclosed colonies. The illustration on page 57 shows what happened in one set of colonies; the results were similar for the other four sets. All enclosed colonies had neighbors encroaching on their foraging ranges within 10 days. I concluded that if a colony doesn't meet its neighbor more often than every 10 days, that neighbor will enter its foraging range. The apparent politeness of older colonies, leading them to use foraging trails that avoid their neighbors' trails, is thus maintained by frequent encounters. Without these encounters, the neighbors will barge in.

As soon as the enclosed colonies were released (by removing the aluminum fences), their neighbors retreated. Being the current occupant of a foraging area doesn't give a neighbor any advantage. The intruding neighbors were evicted. When the enclosed colonies were released, there was fighting among all five sets of colonies. After a few days, conflicts were resolved, and some of the enclosed colonies seemed to reclaim larger areas than they had formerly occupied (see page 64).

To find out if colony age and size alters the way a colony responds to a neighbor's absence, I repeated the enclosure experiments with colonies of different ages. There were three age classes of colonies: 1-year-olds, 3-to-4-year-olds, and colonies 5 years or older. Enclosed colonies were either old (> = 5) or intermediate (3 to 4). Neighbors of the enclosed colony were either young (1 year), intermediate (3 to 4 years), or old (> = 5). There were three pairs of each age

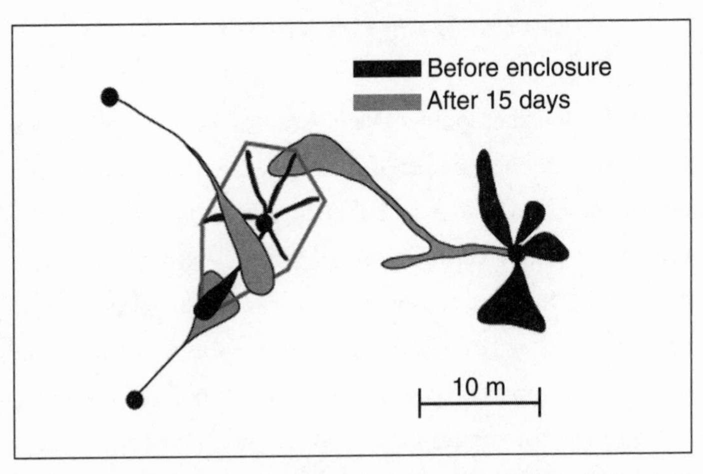

These are the foraging trails of an enclosed colony and its neighbors. Nest mounds are shown with filled circles. The enclosure went around the outer edge of the nest mound. The polygon surrounds the trails of the enclosed colony the day before it was enclosed. When a colony was enclosed, its neighbors entered its foraging range.

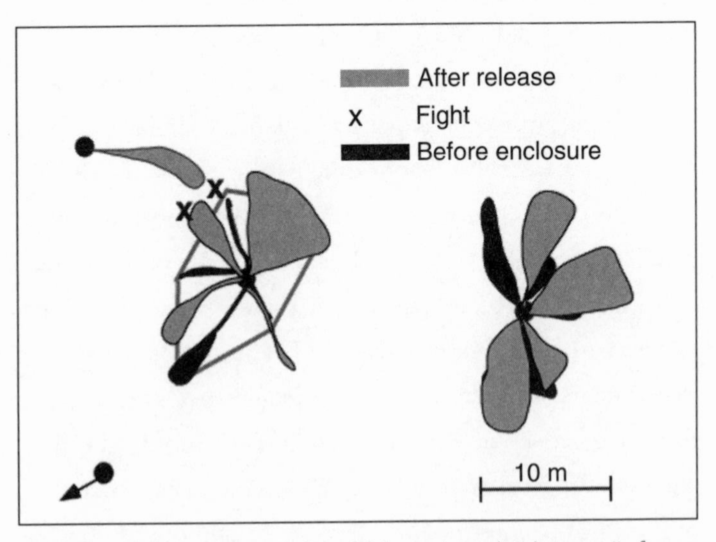

When the enclosed colony was released, it began immediately to use its former foraging area, and its neighbors retreated. There was still some fighting, shown with an X, at the edges of the newly released colony's foraging range.

combination. Colonies were enclosed for 15 days, and foraging maps were made before, during, and after enclosure.

Neighbors of all ages reacted the same way to the disappearance of the enclosed colony. They turned their trails toward the enclosed colony, and they extended their foraging areas. But when the enclosed colonies were released, while all neighbors retracted their trails somewhat, there was a range of responses. The smallest, 1-year-old colonies actually retreated to a smaller area than they had held before their neighbor disappeared. In contrast, the 3- to 4-year-olds tended to keep the area they had gained. This was true even if the released colony was larger and older. Brute force, or total colony size, does not determine the outcome of encounters; 3-to-4-year-old colonies are more persistent and more likely to win a conflict over newly acquired foraging area.

How neighbors reacted when a colony was enclosed did not seem much affected by the age of the enclosed colony. Neighbors did not seem to remember any previous assessment of the enclosed colony's strength; they were not any more respectful of an absent large colony than of an absent small one.

Taken together, the foraging maps of undisturbed colonies, the bait experiment, and the enclosure experiment suggested that colonies perpetually adjust their foraging areas, depending on where their neighbors forage. This process seems to change as a colony grows older. Older colonies avoid trails that meet those of their neighbors, but younger colonies continue in prolonged encounters with one another.

So far I had pulled some stories out of my observations, and done experiments that seemed to confirm these stories. I still wanted to be sure that in the absence of any interference from me, colonies behave the way that the experiments suggested they do. The next step, then, was to watch many colonies, without interfering, and find out how much their interaction with neighbors could predict where they foraged. We watched all the foraging trails of 34 colonies for 17 days, recording the location of all encounters between colonies. An encounter area was a region of about 1 m^2 where the foraging trails of two colonies met. We measured the locations of encounters, using a tape measure and compass, from small metal pegs we already had in the ground as markers for measuring the locations of nests.

Encounters between colonies are very frequent. In the course of 17 days we saw 1,090 encounters, or 1.9 encounters per colony per day. Our 17 days of observation were at the height of the summer foraging season. The more foragers a colony sends out, the greater the distance they cover. On days when all colonies have large numbers of foragers out and thus large foraging areas, overlap between colonies is more likely. In other seasons, when foraging is less intense, the rate of encounters is probably lower.

How likely one colony is to meet another depends on their

neighborhood. The closer two colonies are, the more likely they are to meet; the probability of meeting decreases by a factor of 0.16 for every 10 m of distance between them. Also, the larger a colony, the more likely it is to meet its neighbors. On the other hand, colonies are less likely to meet their neighbors if they met on the previous day, but the probability depends on the age of the neighbor. Young neighbors, 1 or 2 years old, are even less likely than 5-year-olds to return to the site of an encounter on the previous day. But 3-to-4-year-old colonies are about 2.5 times as likely as 5-year-olds to return to the site of an encounter on the previous day.

The more often a pair of colonies meets, the more distant from each other their encounters become. It is as though they attempt to avoid meeting, but can't avoid it; both colonies take routes that lead each one away from the site of the original encounter, but living so close together, eventually they are bound to meet somewhere else.

All of the ways I looked at neighbor relations—seed baits, enclosure experiments, and watching undisturbed colonies—lead to the same conclusion. Very young, small colonies (<= 2 yr.) and mature, older colonies (>= 5 yr.) avoid foraging at sites where they recently met a neighbor. A colony of intermediate age, at the steepest part of its growth curve and a year or two before it begins to reproduce, behaves very differently. It will persist in returning to the site of an encounter, even if this means fighting.

During a colony's life cycle, its response to its neighbors

changes. Why? Neighbors affect the chances a colony can get started, survive, and reproduce.

GETTING TO KNOW THE NEIGHBORS

In most social insects, a worker can recognize another worker from its own nest. Workers seem to acquire a colony-specific odor by living in the colony, perhaps from a substance that they spread on each other in mutual grooming. An ant will be accepted as a colony member if it is put into the nest as a pupa or younger, even if it is of another species.

Ants have well-developed systems for distinguishing "self," the same colony, from "other," all other colonies. I wondered if harvester ants could make a further distinction, between two kinds of "other": ants from neighboring nests, and ones from nests so far away that foragers or patrollers would never meet. Harvester ant neighborhoods set the stage for long-term relationships: the same colonies may live next to each other for 15 or 20 years. Might ants get to know their neighbors? Birds do, learning the songs of particular neighbors so well that a tape recording of a neighbor's song, played from the wrong direction, will send a bird into great confusion. The analogous experiment with ants would be to present them somehow with the scents of neighbors and strangers, and see if they react differently. There are two problems with this plan. The technical one is that we don't know how to extract the colony-specific odor from ants in a form we can easily spread around. The deeper problem is that we don't know exactly when and where ants perceive their neighbor's scent. One common method is to put ants together in a box and see how they react; they might fight or react in a hostile way which would indicate they perceive each other as members of different colonies. But ants in a box may just be reacting strongly to the strange experience of being in a box. To get around these problems, I staged encounters between the foragers of one colony and ants from a neighboring one, and compared the reaction with that of foragers exposed to strangers.

My experiment employed groups of three focal colonies: one exposed to neighbors, one to strangers, and one undisturbed as a control. I measured the foraging rate of all three focal colonies, to see if the colonies exposed to neighbors or strangers would differ. I collected 10 ants from a neighboring colony and put them down near the foraging trail of the first focal colony. I collected 10 ants from a distant colony (more than 150 m away) and put them down near the foraging trail of the second focal colony. Ants forage, and colonies move, only over distances much smaller than 150 m, so ants from such distant colonies ordinarily would never meet. Ants of the groups of 10 that were put down near foraging trails usually ran around in circles for a while and then went away. The third focal colony had no ants put down near it, and was used as a control for the effect of current temperature on foraging; everything else being equal, there is more foraging when it is warmer. We counted the numbers foraging every 5 minutes for 40 minutes, with three people counting the three colonies simultaneously. We repeated this experiment nine times with seven colonies (not counting the colonies from which ants were collected).

Harvester ants do distinguish neighbors from strangers. While both the colonies exposed to neighbor ants and the colonies exposed to stranger ants foraged less than the undisturbed colony, there was a significant difference between the two. Foraging decreased more when ants were exposed to neighbors than when they were exposed to strangers.

These results are consistent with the observations of mutual avoidance in older colonies. When foragers from one colony encounter foragers of a neighboring colony, foraging slows down on that trail. From one day to the next, the outcome would be that the foraging trails leading to the place of the encounter are used less and less, and trails in other directions are used more.

The stranger ants were alarmed and probably created some disturbance because they were alarmed, and this upset accounts for the decrease in foraging, relative to the controls, by colonies exposed to strangers. But the colony's reaction to stranger ants, carrying a scent presumably never encountered before, was not as strong as the reaction to neighbors. A stranger ant is a lost ant, and it poses no threat. A neighbor ant belongs to a colony that may be living nearby and competing for foraging area for the next fifteen or twenty years.

If ants from some colony recognize the ants of a neighboring colony, they must meet often enough to learn and remember the colony-specific odor of their neighbors. How often remains an open and intriguing question. Perhaps certain foragers consistently use certain trails, as they do, for example, in the huge red wood-ant colonies of northern Europe. If so, then there might be a particular group of trail A ants in one colony that tend to meet a particular group of trail C ants from a neighboring colony. If the ants' memory is long, or encounters are frequent, then most ants from the two trails should recognize each other. But if memory is short, or encounters are infrequent, or individual foragers mix randomly on all trails, then the ability to recognize would be more rare.

A simple experiment showed that foragers are not completely committed to particular trails. I marked ants one day on a particular trail. The next day, I put out food on a second trail. More ants appeared on the trail with the seeds, including ants that I had marked on the first trail the day before. When food availability changes, foragers switch trails. The day-to-day turnover in individual use of certain trails must depend in part on how the distribution of food changes from one day to the next.

Mark Brown came to work with me in Arizona when he was an undergraduate at Oxford, got hooked on ants, and then came to work with me at Stanford for his Ph.D. research. He and I wondered whether certain foragers might specialize in interaction with neighbors. If so, these ants might come to learn the odor of neighboring colonies, and when they meet neighbors, might somehow initiate the withdrawal of the rest of their own colony's foragers. In the laboratory, we set up pairs of colonies so that their foragers would enter opposite sides of an arena divided by a partition with a little ant-size sliding door in it. We opened the door, allowed some ants to enter the foraging arena of the other colony, and marked all the ants that interacted with those of the other colony. An interaction consisted of a brief antennal contact or a fight. We then tested whether some ants were more likely to interact than others, and whether some were more likely to fight.

No ants were devoted to diplomacy; all ants were equally likely to encounter ants of the other colony. This finding makes it difficult to explain how neighbor-stranger discrimination can arise. Following foragers in the field to measure how often each one meets an ant of a neighboring colony, I found that the probability of such a meeting is small, about 0.06 meetings per day. If an ant forages for about 30 days, it will meet only 1.8 ants (0.06 × 30) of a neighboring colony. It's hard to imagine this is enough for many ants to learn the colony-specific odor of a neighbor. Ants may also communicate somehow to their nestmates about the odor of the neighbors they meet.

Some foragers specialize in fighting. The same few ants tended to do most of the fighting. This job is not as dangerous as it sounds, because surprisingly few ants die in fights. We measured the risk in the field by watching 133 fights. Most fights last less than a minute. Only 21 percent ended in injury or death for any of the participants. The existence of specialized fighters might explain why fighting is so pervasive on some days, and rare on others. Perhaps on days when most of the foragers are active outside the nest, the specialized fighters are likely to be active as well, and once they are out, itching for trouble, fights with neighbors are more likely to occur.

Relations between colonies are elaborate: ants recognize their neighbors, colonies adjust their trails around the trails of their neighbors, and colonies develop new diplomatic maneuvers as they grow older and larger. In the dialogue between colonies, the tone is set by the ages of the colonies in the neighborhood. Neighbors are competing for foraging area; what one colony does not use, another will. What is at stake in this competition? The next chapter explores why neighborhood politics matter.

4

A FOREST OF ANT COLONIES

•

We both drank the wine.
"What do you think of the war really?" I asked.
"I think it is stupid."
"Who do you think will win it?"
"Italy."
"Why?"
"They are a younger nation."
"Do younger nations always win wars?"
"They are apt to for a time."
"Then what happens?"
"They become older nations."

Ernest Hemingway, *A Farewell to Arms*

•

BIRTH AND DEATH

A population of harvester ant colonies is like a forest in which the trees compete for light. Large trees shade the ground, making it difficult for seedlings to survive nearby. An ant colony collects food in the surrounding area, and large colonies may make it difficult for new ones to be founded nearby. I worked on this question with Alan Kulig (then a graduate student in statistics and now the investment manager for a large insurance company). We tested whether the probability that a one-year-old colony appears in a given location depends on the ages and sizes of neighboring colonies. We found that, like trees, large colonies make it difficult for new ones to survive nearby. New colonies are more likely to appear near young, small colonies than near old, large ones.

Few newly mated queens survive to dig new nests the day after the mating flight; tens of thousands of alates may be at the flight, but I have never seen more than about a thousand new nests the next day. Vulnerable newly mated queens get eaten by birds, lizards, and other ants. A queen who lands near a large harvester ant colony is more likely to be found by a worker and dragged away before she can dig a nest than a queen who lands near a small colony. Then neighbors make life difficult for queens who do manage to make a nest and hide in it. Of the 1,000 new nests I see each year scattered all over the site the day after the mating flight, I find only 20 to 50 as one-year-olds the following summer. These tend to be clumped in neighborhoods of small, young colonies. A founding colony with large neighbors probably cannot get enough food to support the first few batches of workers, and by the next summer, the first workers and the queen will have died.

Competition for food explains why there are crowded clusters of colonies of similar age. In a given year, all of the new colonies close to small ones are most likely to survive, leading to neighborhoods of small colonies. Eventually such clusters will grow into clusters of large colonies years later, and these will muscle out new colonies that try to enter the neighborhood.

Once a colony is established, its neighbors rarely starve it to death. Alan Kulig and I tested whether neighbors affect the probability that a colony dies within a given year. Mortality is low for colonies two years or older, ranging from 4 to 11 percent, and in most years the threat of dying is no worse in a crowded neighborhood than in a sparse one. In some years colonies with young, small neighbors are a bit more likely to survive. It makes no difference where on the site the colony is located. What predicts mortality far better than anything about the neighborhood is colony age: harvester ant colonies tend to die when they reach a graceful old age.

THE NEXT GENERATION

Competition for foraging area can have a strong effect on reproduction. One measure of reproductive success is the number of alates a colony can produce. This is not an entirely satisfactory measure, because the mortality of alates is so high. Everything else being equal, a colony that produces 100 alates probably has a better chance of having a daughter colony established in the next generation than a colony that produces only 10. But everything else may not be equal and indeed, usually isn't. Perhaps a colony does best to make a small number of females whose good health increases the chances they will escape predators, quickly construct a superior nest, or raise unusually robust foragers. Knowing little about what makes a female alate likely to be one of the lucky few to survive, or about what makes a male likely to mate at all, we have settled so far for counts of alates as a measure of reproductive potential.

We try to catch the alates as they come out for the mating flight, just long enough to count them, and then let them go so they can mate. This procedure is much harder for us than catching them in a way that kills them, so we could count them at our leisure—but I try to minimize my interference with the dynamics of the population. Few alates survive to found new colonies, so killing them all off some year would drastically alter who contributes to the population in future years.

Over several years Diane Wagner and I designed a pretty good alate trap. Diane came to work in my lab as a postdoctoral researcher; now she is a professor at the University of Nevada, Las Vegas, studying ant ecology. The trap consists of two cones, one huge and one small, made of aluminum screening. The large cone is about 1 meter high, its base about 1 m across, so that sitting with its base on the ground, it encircles the nest mound. Fitted to the base, there is a cylindrical support for the cone to sit on, made of hardware cloth of mesh coarse enough for the workers to pass freely as they enter or leave the nest. The top of the larger cone doesn't come to a point; its end is cut off. Around this top end of the larger cone sits the smaller cone, about 10 cm high, like a stiff Santa's hat. The principle is the same as that of a catfish trap. The alates fly up through the cut-off end of the larger cone and into the smaller cone, and then cannot find their way back down. On the day of the mating flight, we come along just as the alates have gone into the cone, count them, lift off the smaller cone, and let the ants fly off to the mating site.

In the first year, we made about 40 of the traps. Then we decided to repeat the measurements on a larger scale. As that decision was congealing, someone showed me a letter from a local Boy Scout, Josh Olivo, explaining that his troop could earn merit badges by contributing to an ecological research project. Josh and his friends ended up making almost 200 traps, with a lot of help from Josh's father. Their merit badges were well earned, and Josh's father deserves a medal. Diane Wagner and Mark Brown loaded the traps from Josh's garage in San Jose, California, into a U-Haul and drove them to Arizona.

Since we don't know when the mating flight will be, the traps have to be in place for many days, even weeks. In the second summer of the study, when we put out traps on 180 colonies, the site looked like a parking lot for a large fleet of dilapidated spaceships. We tried fastening the smaller cone to the larger one with duct tape, which led to many days wasted chasing after the cones that came off when the duct tape dried out, and endless hours replacing it with yet more duct tape. A much better solution was a band of white soft synthetic cotton around the inner edge of the small cone; this trapped the alates and the cones could just be stapled on. The synthetic was impervious to rain and wind. Probably it was not invented for trapping harvester ant reproductives—it looks like it could be snow in a store window's Christmas display—but it worked very well.

We trapped reproductives in two climatically different years and got different results. The first year, 1995, was a dry year following many dry ones. Crowded colonies produced fewer alates. The second year, 1997, was a wet year following another wet year. Neighborhood density had no effect on reproduction. Some colonies were measured in both years and the colonies that had done well the first time also did well the second. In each year, surprising numbers of colonies, about 25 percent, did not reproduce at all. Colonies tend to make more alates as they grow older. We caught some alates to weigh and found no relation between their weights and the numbers produced; there is no tradeoff leading some colonies to produce fewer, heavier alates and others to produce more light ones.

In wet years, there is more food available to harvester ants, because plants make more seeds. So when it is dry, and resources are scarce, competition for food is probably more intense. Then colonies are under more pressure from their neighbors, and crowded colonies make fewer alates. In a good year, by contrast, neighbors are less important.

Competition for foraging area affects whether a colony can get started at all, but also the number of alates it can make. Very few queens manage to establish colonies, and although most survive, surprisingly few manage to reproduce. It seems a few colonies produce most of the population's new colonies, year after year. How does a colony get to be one of the lucky few that reproduces?

COLONY BIOGRAPHIES

Some ant colonies live long and fruitful lives; others struggle along unable to get rich enough or large enough to reproduce. For almost 10 years the most crowded patch of the study site is the "inner city" area described above, packed with colonies mostly founded in 1988. Now they are much older than five and should have begun to reproduce years ago. But most of the colonies in this area look much smaller than five, and few have ever produced any alates.

There may be a link between a colony's size and its ability to reproduce. At a given age, some colonies are larger than others. This variation cannot be measured directly while monitoring the population, because to count the ants in a colony one has to dig it up and probably destroy it. But it is obvious that colonies of the same age vary in the numbers of foragers and numbers of other ants working outside the nests. And as we saw in chapter 2, the excavations showed that numbers working outside the nest are correlated, loosely, with the total numbers of ants inside a colony. Suppose a colony needs some threshold amount of food before it will begin to make alates. A queen weighs about ten times as much as a worker, so the colony must use more food to produce alates than workers. Colonies with more workers may have larger stores of food, because they have more foragers to get food and more ants inside the nest to process and distribute the food, transforming food into ants ever more effectively as the colony grows. Thus a colony facing severe competition from its neighbors may never grow large enough to reproduce. We can imagine what selection pressures could operate in this system. Colonies that somehow find the means to grow larger in the face of competition, might contribute more colonies to the next generation. Over evolutionary time, if the colony's behavior is heritable—and we don't know if it is—such behavior might increase in frequency.

Colony behavior is linked to colony life history. The 3-to-4-year-old colonies are more persistent in territorial conflict than older ones, from 5 to 15 years old. One proximate reason for this may be the demands of colony growth. Foragers in a growing colony may respond to the demands of many little larval mouths to feed. A 3-to-4-year-old colony is in the steepest part of its growth curve. Adult workers apparently eat little; most of a colony's food goes to its larvae. In a 3-to-4-year-old colony, the workers must collect and process the food necessary to make the colony larger next year. Since workers live only a year, the colony must recreate itself each year. The 4,000 ants of a 3-year-old colony have to feed 6,000 larvae to make the 4-year-old colony. For a colony 5 years or older, already at its adult size, the situation is different. For every new ant to feed, there is already an ant there to help feed it. The colony must produce 10,000 workers each year to maintain its size of 10,000 workers—but it has 10,000 workers to collect and process the food necessary to do this. So the demand for food, per forager, may be greater in the smaller, quickly growing colony. This might make foragers of a small, quickly growing colony more prone to engage in conflict over food than those of a larger one.

THE FUTURE VALUE OF FORAGING AREA

What does a young colony stand to gain from being more persistent and aggressive than its older neighbors? One possibility is that young colonies are willing to fight harder to keep especially valuable regions of foraging area for the future. But we know enough about the distribution of food to rule this answer out. There are no especially valuable regions. During the summer, when colonies are fighting with each other, one area offers more or less the same kinds of seeds as another, all mostly distributed by wind and flooding. No consistently abundant patches of food persist day after day. Seeds carried by wind and water land in small depressions in the soil, or collect at the base of plants. The shape of the soil surface changes constantly. A place that collects many seeds today may not do so tomorrow, and there is no guarantee it will do so year after year. This uncertainty means the ants cannot evaluate, or base their fighting behavior on, the future value of a particular spot.

Whether the young colonies are claiming foraging area for the future is another question. If a colony tends to keep the same foraging area year after year, then a 3-to-4-year-old colony could be staking a claim to an area it will keep for the rest of its life. The older colony might be less persistent if it already has staked out its permanent foraging area, and needs only encounter its neighbors often enough to keep them at bay.

To find out how long a colony holds on to its foraging area, I measured how much of the foraging area used one year was used the preceding year. Suppose a colony keeps exactly the same foraging area from year to year, or keeps it and expands around the core area. Then each year's foraging area would overlap completely the foraging area of the preceding year. I used foraging maps made for the same colonies in successive years, a total of 404 maps for 88 colonies. I had to take into account the variation from day to day in a colony's foraging area, pooling the data from different days in the same year. I simply combined all the foraging areas of a given colony in a given year into a single cumulative foraging map for that year. Then I measured the extent of overlap in the cumulative foraging maps for two successive years. Two different measures were calculated; in one of them, regions within the cumulative foraging area were weighted according to the number of days they were used. In the second, all regions were weighted equally. The choice of weighting method did not affect the main result, which means that day-to-day variation did not make much difference.

On average, only half of a colony's foraging area from a given year was used the previous year. Overlap from one year to the next was significantly greater than it would be if foraging areas were located randomly with respect to their location the previous year. But year-to-year overlap was not anywhere close to 100 percent, as it would be if foraging areas remained constant or expanded from a central core.

Using the same data, I also examined whether a colony's foraging area grows in size as the colony grows older and larger. Surprisingly, it does not. There was no significant difference in the total area occupied in a given year by small 2-year-old colonies from that occupied by much larger colonies aged 5 years or more.

If larger colonies do not occupy larger foraging areas, why are they more likely to meet their neighbors than smaller ones? First, area was measured from foraging maps, showing area occupied by ants above a threshold density. The probability that two colonies meet depends on the probability that ants from each colony will be in the same place at the same time. The more often a colony sends foragers to a place, the more likely it is to meet ants of a neighboring colony. Larger colonies have more foragers than smaller ones. This means that any place in a large colony's foraging range is probably visited more often by foragers than any place in a smaller colony's range. Large colonies have more ants available to meet neighbors than do small colonies. Second, larger colonies forage greater distances than smaller ones. The farther a foraging trail extends toward a neighbor's nest, the more likely are ants of the two colonies to meet. Smaller colonies can occupy the same total area as larger ones by having many short trails instead of a few long ones.

If 3-to-4-year-old colonies are not fighting for an area they will hold on to in the future, do they gain anything from returning to the site of an encounter with a neighbor? The other side of this question, why it might be worthwhile for a colony to persist in encounters with neighbors, is what the encounters cost a colony. If they do not cost very much, then the 3-to-4 year-old colony does not have much to lose.

Encounters might have two kinds of costs for a colony. The first is death or injury due to fighting. The second is foraging time wasted; while an ant is interacting with an ant from another colony, it cannot search for food or carry it back to the nest. The amount of food a colony obtains is limited by the number of ants actively foraging and by the constraints, set by high afternoon temperatures, on how many hours the colony's foragers can be active. If the foragers spend a lot of time interacting with neighbors, they will have less time to bring in food.

I measured the costs of encounters by comparing the duration of an ant's foraging trips in two colonies simultaneously. One colony was engaged in an encounter; one of its foraging trails met one of a neighbor's. The other colony had no encounters with any neighbors. The duration of a foraging trip depends on the speed at which the forager moves, which in turn depends on temperature; ants, like other animals that cannot regulate their internal temperatures, move faster when their muscles are warmer. The duration of the trip also depends on how far the forager goes before it begins to search for seeds, and because older colonies tend to have longer foraging trails, the duration might be correlated with colony age. To account for these two sources of variation, temperature and colony age, we followed foragers in two colonies of the same age, at the same time. Each forager was followed from the time it left the nest until it returned.

Few of the ants foraging on trails that intersected one of a neighboring colony actually met an ant of the neighboring colony. Of the direct encounters between non-nestmate ants, few led to fighting. Of the fights, few resulted in injury or death. This means that the costs of encounters are surprisingly low. There was no significant difference in the duration of foraging trips along trails meeting those of a neighbor and those in which no such interaction occurred. Foragers did not spend much time involved in interactions.

It seems there is little at stake in behavioral interactions between colonies. Whether two colonies meet or not, and even whether they fight or not, does not much affect how much food they each get. The rate of encounters between two neighbors is important because it reflects the intensity of their competition for food. An encounter occurs when two colonies search the same place on the same day, but over the season colonies tend to search the same places repeatedly. When neighboring colonies meet, it means their foraging areas overlap, even when they search the same area on different days. Seeds are renewed very slowly, so food taken by one colony on one day probably means less food for another on subsequent days. When a colony has many encounters with a neighbor, that neighbor is a strong competitor for food.

ULTIMATE SUCCESS: GETTING LARGE ENOUGH TO REPRODUCE

Competition for food matters a great deal to an ant colony, because it affects whether a colony can get started at all and how much it can reproduce. Evolutionary questions about harvester ant behavior center around foraging, because foraging behavior is so important for a colony's survival and reproduction. Foraging is linked to the other tasks a colony performs, embedded in a daily round of activities. As the colony's circumstances change, the daily round stretches and constricts, like a lump of clay, but preserves the basic pattern. These adjustments arise from the simple decisions of workers. Small shifts in the probability that a worker will turn one way or another, take up one task or stay inside the nest, all add up to a predictable, imperfect but adequate, response by the colony to a changing world. Over the millions of years that ant-colony behavior has been evolving, when has a colony's foraging behavior really made a difference to its survival and reproduction?

Colonies vary in foraging behavior. Some colonies forage on more days than others, and this tendency persists year after year. We do not know whether variation in foraging behavior is heritable. If it is, then natural selection may be shaping colony organization through its effects on foraging behavior. Colonies in more crowded areas produce fewer reproductives. The amount of food a colony has probably limits the number of reproductives it can make. Colonies that forage more, or forage better, may be the parents of more colonies than colonies that forage less. If the offspring of good foragers tend to be good foragers, then over many generations natural selection will increase the proportion of good foragers in the population.

Almost nothing is known about the inheritance of behavior in ant colonies. It is rare even to identify the parents of an ant colony. The parent colonies of a new colony are the one that produced the queen and the one that produced the male or males that mated with her. Because virgin queens and males fly into a mating flight, the males die, and the newly mated queens fly off probably at random, it is virtually impossible to track ants from the parent colony into the mating flight and out again. Any scheme we know of for marking queens would take an hour or two and prevent them from getting to the mating flight in the first place. Even if marked queens could be sent into the mating flight, great luck would be required ever to see them again; if they fly off at random, we would have to search the entire twenty-five acres very thoroughly for the marked ones. A more promising technique is to track the parents genetically. We are now working to identify some genetic markers we can use to recognize the markers of the parent colonies in ants of the offspring colony. Then we hope to be able to ask whether offspring colonies resemble parent ones in foraging behavior.

Natural selection may have acted in the past to shape the behavior that we now observe. It may not be happening now, or it may be occurring but too weakly for us to notice. It is fairly easy to make a good argument for why certain forms of foraging behavior should be selected for, but a good story does not guarantee that we will find selection when we look for it. We are ready to look for evolution in action when the genetic techniques are available to identify offspring colonies. If natural selection currently favors more and better foraging, then we will find heritable variation in foraging behavior, and the best foragers will be producing the most new colonies.

I see little that seems efficient about the ways that ants forage or interact with their neighbors. Their behavior clearly works; there are a lot of harvester ant colonies out there, more every year. I deeply admire their harvester-antness, the richness of their responses to a world so alien to me, but I am never struck by their perfection. Perhaps natural selection is not acting on harvester ant behavior, or perhaps it is acting too slowly for me to see it, or perhaps when I understand more about how colony behavior varies, I will be able to see why some behavior is especially likely to promote reproductive success.

It is always easier for me to think about the evolution of colony behavior when I am in my office far away from Arizona. When I am standing in the field of ant colonies, so much seems to impinge on the ants. There are so many links to the plants, the other animals, the rain and soil, that it is hard to believe that we could ever guess the direction that selection is taking. The story I have told so far pits one colony against another in the search for seeds. But the ants live with many other species of seed-eating ants, and probably those relations, like many others, make a difference.

When we ask why 3-to-4-year-olds are so persistent in encounters with neighbors, a proximal answer comes more easily than an evolutionary, ultimate one. Perhaps growing colonies are hungrier, and because fighting doesn't cost much, any possibility of getting more food is worth the minor inconvenience of meeting the neighbor. If the food supply of older colonies is more secure, because they have more food stored up, the minor hassle of encounters may be enough to make avoidance preferable.

Colonies that manage to grow large enough by age four or five may be able to start reproducing sooner. Considering how few alates a colony makes and how few alates survive to form new colonies, each year's reproductive effort contributes enormously to a colony's small chances to make an offspring colony.

A population of ant colonies is like a forest. Each colony, planted in a neighborhood, uses its foragers to collect the resources it needs. As the branches of neighboring trees jostle for light, so the foragers of neighboring colonies divide up the seeds on the ground. The forest shapes the trees as they grow, as pressure from its neighbors shapes the behavior of an ant colony. To understand how ant behavior evolves, we will need to put all this together, from the bumbling reactions of workers, to the local struggles of neighorhoods to divide up scarce food, to shifts, over many generations of colonies, in the ways that colonies work.

Here is the connection between neighbor relations and the internal organization of a colony. Colony organization determines when ants forage, and how many ants forage. How much foraging a colony can accomplish determines its chances to reproduce. Any behavior that makes a colony more likely to be one of the few that reproduces could, if heritable, be selected. This leads to a question about the internal organization of a colony: What determines how much a colony forages?

5

IN THE SOCIETY OF ANTS

•

"—then you don't like *all* insects?" the Gnat went on....

"I like them when they can talk," Alice said. "None of them ever talk, where *I* come from."

"What sort of insects do you rejoice in, where *you* come from?" the Gnat inquired.

"I don't *rejoice* in insects at all," Alice explained, "because I'm rather afraid of them—at least the large kinds. But I can tell you the names of some of them."

"Of course they answer to their names?" the Gnat remarked carelessly.

"I never knew them do it."

"What's the use of their having names," the Gnat said, "if they won't answer to them?"

"No use to *them*," said Alice; "but it's useful to the people that name them, I suppose. If not, why do things have names at all?"

"I can't say," the Gnat replied.

Lewis Carroll, *Through the Looking Glass*

•

If only ants could talk, our work would be much easier. I wouldn't expect an ant to explain to me how its colony functions (though of course if ants could talk such explanations would naturally follow). But it would be really helpful if an ant could provide a running commentary on what it notices. We tend to impose a social structure of our own invention onto an alien society. How do we find out which features of ant society are important for the ants?

In harvester ants, where colonies are founded, how long they survive, and how much they reproduce, all arise from the relations among neighboring colonies, which influence where colonies forage. Patterns of foraging behavior are part of the network of colony organization, which encompasses all of the colony's tasks. This organization emerges from the ways that individuals respond to their environment and to their interactions with other ants. The responses of ants are behavioral patterns that spring out of a dense web of physiological pathways, and the ant's physiology, its living body, in turn arises from the mysterious interplay of genes and environment as ants develop and go about their lives.

It would be a mistake, however, to believe that the physiology of ants determines the dynamics of populations. Physiological, social, and ecological processes all operate simultaneously and none is more important or fundamental than another. Linking levels of organization is central to any study of social behavior. For humans and other social animals, an individual's behavior is always embedded in a social world.

One example of the force exerted by social context is an ant's response to a chemical cue. Most ants, harvester ants included, have very poor vision. They perceive the world, and each other, mostly through odors. Ants have a large number of glands, up to 14, and each secretes a different chemical product. Some are dispersed into the air; others are smeared on the ground or tree or whatever surface the ant is walking on. Ants perceive these chemicals with their antennae. In some ant species, ants that find a food source lay down a chemical trail as they return to the nest. Other ants at the nest can follow this chemical trail back to the food.

In the early days of research on chemical communication in ants, it was hoped that ant behavior could be explained by finding the pheromones to which each species of ant would respond, and cataloguing the response of each species to each chemical. But there is no one-to-one correspondence between a chemical and a response. Just as the same word can have different meanings in different situations—think of the many different tones and meanings with which someone could say the word "mother" or "yes"—so the same chemical cue can elicit different responses in different social situations.

In 1958, E. O. Wilson and colleagues reported that oleic acid is a "necrophoric" pheromone for harvester ants. Many ant species pile their refuse in a single heap or midden. Wilson concluded that oleic acid elicits the necrophoric response that, in turn, leads to the existence of middens. That is, when an ant encounters an object that smells of oleic acid, the object is taken to the midden. As a dramatic example, the Wilson team treated live ants with oleic acid, and later reported that the ants were carried "live and kicking" to the midden.

I repeated this experiment using not live ants, but pieces of filter paper on which I put a drop of oleic acid. I used filter paper because Wilson had kindly informed me that the live ants used in the 1958 experiments were in fact chilled, so that they were curled up and inert (ants cannot regulate their body temperatures and their muscles will not work when they are cold). If the chemical elicits the behavior, then either curled-up ants or bits of paper should, when doused with oleic acid, elicit the same response from the ants.

Ants showed two very different responses to the bits of paper treated with oleic acid: Either they took them to the midden, or they took them into the nest, as if they were food. There was a pattern to these responses. When ants were foraging, paper treated with oleic acid was taken as food and brought into the nest. This behavior is not surprising, because oleic acid is present in many of the seeds that harvester ants eat. Ants collect seeds covered with so much debris and soil that some strong chemical cues must distinguish seeds. When ants were doing midden work, though, a bit of paper treated with oleic acid was taken as refuse and discarded in the midden. The response to a chemical cue varies, depending on what the ant is doing.

When I marked ants according to task, I found that foraging and midden work tend to be done by different individuals. But this was true only under stable conditions. When conditions change, ants switch tasks (see chapter 7). If extra food appears, an ant that was doing midden work will switch tasks to become a forager. This versatility means the response to oleic acid can't be explained by saying that certain individuals, the foragers, respond to oleic acid as food, whereas other ants, the midden workers, respond to oleic acid as refuse. This can't be true because sometimes midden workers become foragers. When ants switch tasks, they must change their response to oleic acid.

To study chemical communication we must find out how response to chemicals varies with social conditions. In the study of animal behavior more generally, to understand the causes or evolution of behavior we must first learn how that behavior is embedded in a social pattern.

Let's try an absurd illustration. Suppose a Martian marketing firm wants to test whether humans prefer pistachio to peppermint ice cream. You are sitting there reading this book. A giant forceps comes out of the sky, picks you up and puts you down in another room with another chair. Tomorrow morning, a tray appears with two dishes, one containing pistachio and one containing peppermint ice cream. Which one you eat may reflect your preference for pistachio or peppermint, but it may also be influenced by how you react to other aspects of the situation, such as your thoughts on how people you know will respond to your disappearance, all of the appointments you have missed, how much you want to get back to reading this book, and so on. All of these considerations may be more complicated for you than for a laboratory rat, or an ant, but the principle still holds. What an animal does in an experimental situation depends on how and where the situation is embedded in its everyday life.

What was the animal doing on the way to the experiment? This question is often swept away under the "everything else being equal," or *ceteris paribus,* argument. If you take lots of samples, the argument runs, chances are the differences among them caused by differences in context will even out, and what will show up is mostly the effect of the experiment itself. This plan assumes that it is easy to obtain many more samples than there are possible situations, which may underestimate the range of experience of the animal. It also assumes a lack of pattern in the experience of the animal. The experimenter dips a net into the world of the animals, scoops some observations out, and assumes that some representative distribution of result-affecting contexts is obtained. That is, if there are 12 states that an animal can be in, all of which might affect the result of the experiment, the *ceteris paribus* argument assumes that there will be enough sampled to obtain all 12.

But many of the conditions influencing the animals are patterned. If most animals are awake at 9 A.M., then a sample at 9 A.M. will sample mostly awake animals, not sleeping ones. Because the sample at 9 A.M. does not capture all the possible conditions, the experimenter will mistakenly conclude that the way animals react when wide awake is the characteristic response of this species. If the experiment were done at midnight, when most animals are sleepy, the results would be completely different.

Temporal patterns of social behavior are well known, and experiments often take them into account. But the general principle can be missed: Social patterns we don't know about may determine how animals respond. It seems to me most efficient and most biological always to assume that there is an underlying pattern, and to find out what it is before the experiment.

For example, a foraging animal responds differently to food from one that is searching for a nest. If an experimenter learns to distinguish foraging and searching behavior, food can be offered to animals only when they are foraging. This rather simple adjustment makes for a better experiment than one in which food is offered to many animals in the blind hope that numbers searching for nests will be balanced by numbers foraging.

The most direct way to investigate animal behavior is to try to see it as a complete pattern, not to take it apart. The more we take it apart, the more work we have to do to put back together the conditions and other kinds of behavior it belongs with. Sometimes it is argued that we have to take nature apart to understand it, because otherwise we are faced with a complicated swirl of events and no way to tackle it. Trying to understand the whole system would be difficult. There is some balance between the paralyzing contemplation of the complexity of everything, and a focus on components that can each be understood separately but are so isolated that they cannot be traced back to see how they fit into the whole system.

How does an ant colony get through the day, and how does it respond to a changing environment? Colonies proceed through a predictable, orderly sequence of tasks each day. But when more ants are needed to perform a task—when the nest mound is flooded and needs repair, or a new food source appears and needs to be collected—ants appear to do the job, and the daily round stretches to accommodate the new conditions. What gets the ants out to do a task at a certain time? Since no one tells the ants what to do, each ant's behavior must depend on very local interactions with other ants and with the world as an ant perceives it.

An ant's experience is mostly chemical and tactile. It is aware only of what reaches its antennae. Ants move around to tune in to the world. If I could know everything that goes on in an ant colony, I would know where every ant has been, and what it found there, and who it met. Of course, I want to know both more and less than that: how the pattern of all the ants' encounters, with each other and with the world, leads to the behavior that we see. Ants meet, and things happen to them, and they make decisions, and somehow it all coalesces into colony behavior. There must be rules that relate an ant's experience to its subsequent action, rules such as: When you find food, pick it up; when you meet danger, run around in circles and send out the alarm odor. Colony behavior is predictable, which means that ant experience must have some regularity. If a colony forages every day, the conditions that evoke foraging must occur each day. So my efforts to understand colony organization come down to two kinds of questions: What happens to ants, and what do they do as a result? An ant experiences only events nearby, so the regularities in an ant's experience depend on the regularities in the ways it moves around. As ants move around, they meet each other. Patterns of ant movement create patterns of encounter between ants, or ant networks. These networks help to determine what ants do. Let us now consider movement patterns and then go on to the question of how ants decide which task to perform.

6

NETWORKS OF ANT PATHS

•

... I am persuaded that the average ant is a sham.... He goes out foraging, he makes a capture ... it is generally something which can be of no sort of use to himself or anybody else; it is usually seven times bigger than it ought to be; he hunts out the awkwardest place to take hold of it; he lifts it bodily up in the air by main force, and starts; not toward home, but in the opposite direction; not calmly and wisely, but with a frantic haste which is wasteful of his strength; he fetches up against a pebble, and instead of going around it, he climbs over it backward dragging his booty after him, tumbles down on the other side, jumps up in a passion, kicks the dust off his clothes, moistens his hands, grabs his property viciously, yanks it this way, then that, shoves it ahead of him a moment, turns tail and lugs it after him another moment, gets madder and madder, then presently hoists it into the air and goes tearing away in an entirely new direction; comes to a weed; it never occurs to him to go around it; no, he must climb it; and he does climb it, dragging his worthless property to the top—which is as bright a thing to do as it would be for me to carry a sack of flour from Heidelberg to Paris by way of Strasbourg steeple.... At the end of half an hour, he fetches up within six inches of the place he started from and lays his burden down; meantime he has been over all the ground for two yards around, and climbed all the weeds and pebbles he came across. Now he wipes the sweat from his brow, strokes his limbs, and then marches aimlessly off, in as violent a hurry as ever.

Mark Twain (from M. Geisman, ed., *The Higher Animals, A Mark Twain Bestiary*)

•

MOVEMENT AND CONTACT

Intuition tells us that how often ants meet each other depends on how many ants are there and how they move around. But how exactly are encounter rates related to path shape? My first approach to this problem was to find a way to characterize path shapes. Although I didn't think I would understand the relation of path shape and interactions completely by just watching ants move, I did think that I would never understand that relation at all without being able to see the paths. As ants don't leave a colored trail behind them when they move, their path shapes must be transcribed somehow. I thought of an article by Tom Seeley, a professor at Cornell University who studies honeybees. In his study of messenger bees, he traced the paths of bees on the glass covering a flat observation hive, an approach I had long admired because he was able to figure out the function of these messengers by making a picture of the pattern of their behavior. Unfortunately many ants don't do well in flat spaces; in an ant farm they tend to stand around in chambers and not make very interesting paths, so the wax-pencil technique would not work. I first tried filming the harvester ants in the field, and learned that to get a clear enough image to trace ants' paths, it would be much easier to film ants on a flat surface in the lab. I filmed fire ants (*Solenopsis invicta*) in the lab as they searched an unfamiliar arena, then traced the paths by taping transparent paper onto the screen of an editing machine, running the film back over and over to capture paths made simultaneously. From this I learned about the way fire ants explore a novel arena, but I also realized that better technology was needed to record path shape, and that studying path shape and encounter rate case by case, empirically, would not tell me any time soon how the two were related.

A few years later I met Fred Adler, then a graduate student at Cornell and now a professor at the University of Utah. Fred has a delightful way of talking about a mathematical abstraction as if it were a simple mechanical process in three-dimensional space. We worked on a model of the relation of path shape and interaction rate. The model included a third factor: how exhaustively, and how often, a group of individuals covers the ground. Because ants do not see well and respond mostly to chemical information, an ant must be close to anything it perceives. This means that the spatial behavior of ants must solve two problems. Ant colonies need to search the world effectively and to maintain contact with each other so that they can communicate if they find something. The contact may be direct, or there may be a lag. For example, if ants communicate a finding through antennal contact, they must be in the same place at the same time. If they communicate with a volatile pheromone, they must be in contact before the pheromone disperses into the air.

These two problems, searching and maintaining contact,

impose conflicting constraints on the spatial behavior of patrollers. How individual foragers should move in order to search effectively is a large and growing field of research. Such work shows that an effective search method is a convoluted path, close to a random walk, in which a searcher may circle over and over the same ground. But a group of searchers on random walks will not encounter each other often, because each will tend to get stuck in one place. To maintain contact with each other, they should walk in paths approaching straight lines. The problem for networks of patrolling ants thus differs from the problem for individual foragers. The group must continually monitor a region while ants stay in contact with each other.

The model Fred and I constructed described the behavior of a network of patrolling ants, using both simulation and analytical approximation. We assumed that all ants in a group used the same shape of path. Path shape was described in terms of the variation in direction between successive steps along the path. Higher variation in step direction produces a more convoluted path. We wanted to find out how often ants would discover a food source, or an enemy ant, or anything at all worth noticing. In the simulation, such events appeared, scattered randomly throughout the region occupied by the ants, and the duration of the events varied. Ants were considered to disseminate information when they came into contact, that is, within a small distance of each other. We tested how path shape and number of ants affect the efficiency of the network. Efficiency was measured in two ways: what proportion of events were discovered, and the rate at which ants became informed about an event discovered by one of them. Efficiency at covering space depended on the shape of the ants' paths. Efficiency at disseminating information depended on the rate and distribution of encounters.

The effects of number of ants, and of path shape, interact with each other. When there are few ants, straight paths are better than wiggly ones. If a few ants use convoluted paths, the group as a whole tends to miss things, because whenever food appears between two ants going around and around in circles, neither ant is likely to find it. When ant density is high, so that the whole area being searched is crowded with ants, then convoluted paths do not have much effect on efficiency. Even if each ant alone does not cover much ground, wherever food appears some ant will probably already be near enough to find it.

In general, as numbers of ants increase, ants cover more ground and contact each other more often. A larger group of ants is more efficient, both at finding out about events and at disseminating information, than a smaller group. Straighter paths also enhance the rate at which randomly occurring, scattered events are discovered. However, when the number of ants is sufficiently large, network size has a stronger effect than path shape, which thus becomes less important. When ants transfer information to each other, large colonies can patrol efficiently even when individual ants follow circuitous, apparently inefficient paths.

I tested this model with the Argentine ant *Linepithema humile*. This invasive species displaces most native ant species wherever it goes. Our current work suggests their success is

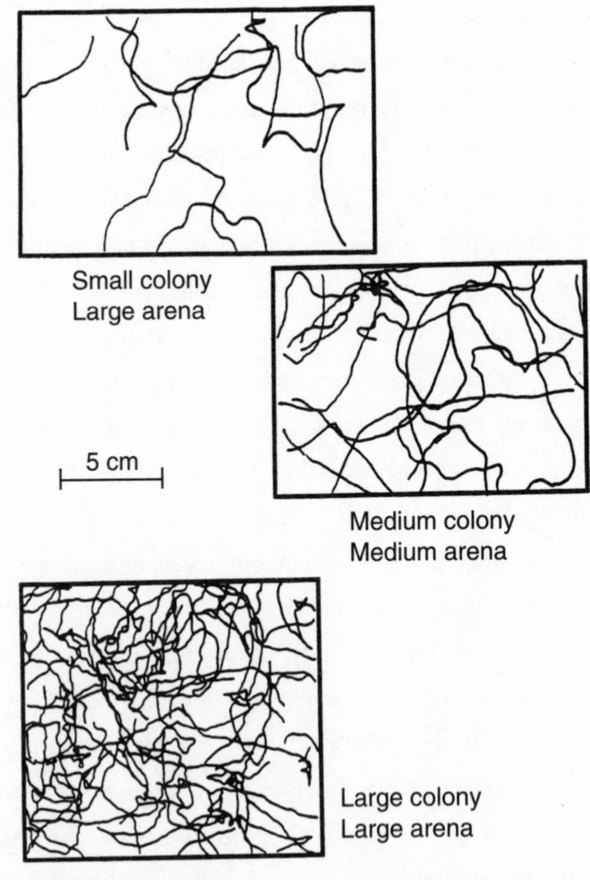

These maps show the paths of Argentine ants at three densities. The fewer the ants (lower density), the straighter the paths.

largely due to their prowess as foragers; they starve out the native competition rather than subduing them by brute force. Instead of laboriously tracing the paths of ants from film, I was able to use image analysis software, developed by Garr Updegraff, to trace the paths of hundreds of ants simultaneously from videotape. The Argentine ants adjust their path shape as a function of worker density in exactly the way the model predicts. When density is high, so that path shape has little effect on searching efficiency, they use convoluted paths. But when density is low, ants use straighter paths.

The paths of ants form an elastic net. When ants are crowded, the net relaxes into twists and curlicues. Perhaps such convoluted paths help to keep the ants near to a food source once they find it; one cookie crumb is likely to be near another. At low density, when ants are scarce, the elastic network of paths stretches out to cover the area more effectively.

Real ants behave as the model predicts, which suggests that the relation of path shape, encounter rate, and searching effectiveness, as described in the model, is important for ants.

COMPARING ENCOUNTER PATTERNS OF ANT SPECIES

The pattern of encounters among ants will depend on how they move around. Ant species differ in characteristic movement patterns; some, such as the red wood-ants of northern

Europe, tend to walk in relatively straight lines. Other ants move in more convoluted paths; for example, the ant *Conomyrma insana* was probably named for its habit of running around in wild circles.

In many ant species, there are daily and hourly temporal patterns of interactions, often in the form of brief antennal contact; best studied are some species in the genus *Leptothorax*. For most species, we do not know the function of such interaction patterns.

I compared encounter patterns of ants in three ecologically very different species, kept in the laboratory in transparent plastic boxes on sheets of paper marked with square grids. I worked with Ric Paul, then on his way to graduate school to study the malaria parasite, and Karen Thorpe, then working as a lab technician. We counted numbers of brief antennal contacts seen in three minutes of observation of each grid square. We made observations several times a day, including feeding time. This work was done in England and one species we observed, *Myrmica rubra,* is a local one. The ants are somewhat reserved, if not timid; they have small colonies and are easily crowded out by other species. The second species, the fire ant *Solenopsis invicta* Buren, is an extremely effective forager that has invaded disturbed habitats in the southern United States. The third species, *Lasius fuliginosus,* which like *M. rubra* occurs throughout northern Europe, has large, long-lived colonies that collect a stable food resource, the honeydew from aphids.

The three species use encounters differently. In the fire ant *S. invicta,* contact rate changes throughout the day. Ants touch antennae more often near food and when food first appears. Fire ants seem to use brief antennal contact to recruit others to food. In *M. rubra,* rates of contact were uniformly low. Perhaps encounter patterns are not important for this species. Contact rates were much higher in *L. fuliginosus* than in the other two species and did not seem to depend on location or the presence of food. The high, steady encounter rates for this species suggest that encounters may have some function not related to the transfer of information about changing external conditions.

A laboratory experiment suggested that in *L. fuliginosus,* contact rate is a cue to density. We tested whether ants react to the rate of encounters they experience with foreign ants of the same species but of a different colony. To do this we set up an experiment to distinguish a response to proportion from a response to number. We reasoned that ants could count to measure number, but to measure proportion they would have to keep track of rate, number of non-nestmates met per unit time. Ants were kept in small, rather crowded boxes, in groups of either 35 or 75 ants. To each host group, we introduced 15 ants of another colony. The rate of contact with non-nestmates depended on the relative proportions of nestmates and non-nestmates. The 15 non-nestmates circulated freely, so each of the 35 ants would tend to encounter one of the 15 non-nestmates at a higher rate than would each of the 75. If the host ants' response depended on the number of alien ants, that is, on the total amount of alien pheromone introduced by the 15 non-nestmates, then both the 35-ant and 75-ant host groups should have had the same response. But the 35-ant groups responded differently from the 75-ant groups. Ants apparently responded to the *rate* of interaction with non-nestmates, that is, to the proportion of non-nestmates present.

When we added ants from another colony, contact rates increased significantly in both the 35- and 75-ant groups. It seems that when an ant meets non-nestmates too often, it increases its rate of contact with all other ants. By stepping up its encounter rate, the ant can discover sooner the proportion of non-nestmates present. Such behavior would be useful for a foraging ant, because a large proportion of encounters with non-nestmates would indicate that the forager had strayed into the territory of another colony. It would also be useful when one colony is invaded by another, as the proportion of non-nestmates would indicate the magnitude of the invasion. How the ants raise their contact rates remains to be discovered. We tested whether the ants run faster when they meet non-nestmates, but they did not.

Ants can use contact rate to assess density, if contact rate depends reliably enough on density. If contact were a consequence of purely random collisions, the total number of pairwise collisions in a group of n particles would be proportional to n^2; that is, adding more particles rapidly increases the total number of collisions. This rapid increase is how numbers of contacts should depend on numbers of ants, if contact arises from random collisions.

We tested whether contact arises from random collisions in *Lasius fuliginosus*. We measured contact rate in a range of densities, by varying both arena size and numbers of ants. When density was low—few ants in a large arena—the ants tended to go to the sides of the arena. This might be because ants like edges, or it might be a way to maintain contact. As the arena grows larger, the distance an ant has to travel around the edge of the arena to reach another ant does not increase nearly as fast as the amount of area the ant would have to cover to search out another in the whole arena. That is, as the arena grows larger the perimeter of the arena increases linearly, adding the length of one edge to the length of another, while the area increases exponentially, multiplying the length of one edge by the length of another. Thus ants can stick together more easily by keeping to the edges. To test whether ants go to the edges of large arenas merely out of a fondness for edges, we put ants in edgeless arenas: on spheres, rubber balls covered with a nylon stocking to help the ants stay on. One sphere was larger than the other, with surface areas that corresponded roughly to the areas of the arenas. If the ants go to the side of the arena because they like edges, they should not cluster on the sphere in the absence of any edges. If they go to the side of the arena to maintain density, they should cluster on the sphere—and they did. (There was no tendency to go to the bottom of the sphere; they seemed quite comfortable staying on the ball, which in curvature was not much different from the large tree branches on which these ants normally travel to collect aphid honeydew.) Maintaining density by clustering is a way to keep contact rate up when numbers of ants are low.

When density is high, *L. fuliginosus* regulates contact rate in the other direction. When there were many ants in a small arena, contact rate did not increase quadratically with density. Instead, it leveled off as density increased. By filming ants walking near each other we could see that an ant could veer off, avoiding contact, when it was about a body length away from another. These ants may be able to see well enough to perceive another ant at that distance.

Contact rate is regulated, as perhaps it should be if it serves as an important cue to ants. At very low densities, ants aggregate and maintain a relatively high contact rate. At very high densities, ants appear to avoid the other ants nearby, so that contact is suppressed. Studies of task allocation suggest why ants might avoid contact when they are crowded, so let us now return to our original question: How do ants, with all their apparent limitations, get things done?

7

SUCCESS WITHOUT MANAGEMENT

•

The Ants toil for no Master
Sufficient to their Need
The daily commerce of the Nest
The storage of their Seed
They meet-and exchange Messages-
But none to none-bows down
They-like God's thoughts-speak each to each-
Without-external-crown.

A. S. Byatt, *Possession*

•

A harvester ant colony performs many tasks: It must collect and distribute food, build a nest, and care for the eggs, larvae, and pupae. It lives in a changing world to which it must respond. When there is a windfall of food, more foragers are needed. When the nest is damaged, extra effort is required for quick repairs. Task allocation is the process that results in certain workers engaged in specific tasks, in numbers appropriate to the current situation.

Task allocation is a solution to a dynamic problem and thus it is a process of continual adjustment. It operates without any central or hierarchical control to direct individual ants into particular tasks. Although "queen" is a term that reminds us of human political systems, the queen is not an authority figure. She lays eggs and is fed and cared for by the workers. She does not decide which worker does what. In a harvester ant colony, many feet of intricate tunnels and chambers and thousands of ants separate the queen, surrounded by interior workers, from the ants working outside the nest and using only the chambers near the surface. It would be physically impossible for the queen to direct every worker's decision about which task to perform and when. The absence of central control may seem counterintuitive, because we are accustomed to hierarchically organized social groups in many aspects of human societies, including universities, business, governments, orchestras, and armies. This mystery underlies the ancient and pervading fascination of social insect societies—how is it that the ant, "having no chief, overseer or ruler, gathers her harvest in the summer . . . to feed the ants in the winter . . ." (Proverbs 6:6).

No ant is able to assess the global needs of the colony, or to count how many workers are engaged in each task and decide how many should be allocated differently. The capacities of individuals are limited. Each worker need make only fairly simple decisions. There is abundant evidence, throughout physics, the social sciences, and biology, that such simple behavior by individuals can lead to predictable patterns in the behavior of groups. It should be possible to explain task allocation in a similar way, as the consequence of simple decisions by individuals.

Investigating task allocation requires two different lines of research. One is to find out what colonies do. The other is to find out how individuals generate the appropriate colony state. An ant is not very smart. It can't make complicated assessments. It probably can't remember anything for very long. Its behavior is based on what it perceives in its immediate environment. In seeking to understand how the behavior of individuals produces the behavior of colonies, the first step is to learn about the outcome, colony behavior. Then we can ask how it is produced.

RELATIONS BETWEEN TASK GROUPS

What determines how much a colony forages? This is a question about the dynamics of foraging, how the number of foragers adjusts to changing conditions. I began to study the dynamics of task allocation by asking how the tasks are related. Does a change in one task lead to a change in another? The numbers engaged in one activity at any time may depend on the numbers engaged in another. We take this interdependence for granted in human societies. It seems obvious that the numbers of people driving cars, working in offices, and eating in restaurants are temporally patterned on daily and weekly time scales, because we are accustomed to the notions of rush hour, business hours, weekends, and mealtimes. We expect the numbers of people engaged in these different activities to be functionally related. If there is a huge accident on the freeway, the time at which large numbers of people begin working in offices might be delayed. On a holiday when most businesses are closed, numbers of people eating lunch in downtown restaurants will decrease.

To learn whether ants of different task groups interact, I made a series of perturbation experiments. I altered the environment of a colony in a way that affected only one activity directly, and looked for an effect on the numbers engaged in other tasks. The results showed that different tasks are not independent. Instead, the behavior of one group of ants influences the behavior of another.

To set up perturbation experiments, I had to find things to do to a colony that would change its behavior, but not drastically. I wanted gently to prod the ants engaged in one task, enough to cause the numbers performing that task to go up or down, but not so much that a widespread disturbance would spread to ants engaged in other tasks. I tried to think up events that were confined to one place, where only the ants of one task group would be. I also tried to create events within the range of things that could naturally happen to a colony, to find out how ants respond to the natural changes in their environments. I spent some weeks hassling the ants with rocks, bits of plastic, and string, sometimes making too strong an impact so that the colony shut down for the day. Eventually I designed spatially local events of the right magnitude to change the numbers of ants engaged in each of three tasks—foraging, patrolling and nest maintenance—without affecting any other task group directly.

One perturbation experiment involved only the foragers directly, causing numbers foraging to decrease. I put small plastic barriers down on the foraging trails, at a place far enough from the nest that only foragers would ever go there. The ants could not go over the barriers, but they could go around them or through them via a few holes in the bottom. The barriers were not impermeable enough to create a disturbance on the foraging trail, but they did create an eddy in the flow of foragers, slowing down foraging traffic on the nest side of the barrier.

Colonies might have reacted to barriers by sending out even more foragers to compensate for the obstacle on the trail, but they did not. The decrease in foraging in the presence of barriers probably mimics colony response to any event that slows down food intake. As already noted, foraging is a costly activity; ants use up water and energy when they move around the hot desert, searching for seeds. A decline in the rate at which foragers bring in food may signal low food availability, when it is not worthwhile to forage.

Other experiments showed that a decline in the rate of incoming food leads to a decrease in the number of foragers coming out. I took away food from foragers returning to the nest, and allowed the foragers to go back to the nest without their food items. I gave the foragers time to calm down, so that they did not appear to be alarmed by the time they went back in. Colonies in which foragers had been deprived foraged at a lower rate, relative to undisturbed ones observed at the same time. When fewer foragers are bringing in food, because of obstacles or a scarcity of food, fewer foragers leave the nest to search for more food.

A second perturbation experiment involved only nest maintenance workers directly, causing numbers performing nest maintenance to increase. Early in the morning, when the nest maintenance workers were first active (see chart on page 33), I put out small piles of toothpicks next to the nest entrance. Within about forty minutes, the nest maintenance workers carried all the toothpicks to the edge of the nest mound and left them there. Ants engaged in other tasks walked over the toothpicks, apparently ignoring them completely. Carrying toothpicks required extra nest maintenance workers. The numbers engaged in nest maintenance work increased, including both the ants carrying toothpicks and those doing nest maintenance as usual, carrying dirt out of the nest.

A third perturbation caused an increase in the numbers of ants patrolling. I suspended with wire a small cardboard cylinder from a heavy wire stake, and planted the stake near the nest entrance. The cylinder wobbled in the wind, and patrollers would climb onto the stake and try to get rid of the offending structure. They were unable to alter it, because they could not cut through the wire or cardboard with their mandibles. Eventually, they ignored it. As a second part of the same experiment, I collected ten ants from a nearby colony of a competing species of seed-eating ant, *Aphaenogaster cockerelli*. Ants of these two species fight over food and nest sites. At the same time that I put out cylinders, I put the ten *Aphaenogaster* workers down on the nest mound of the red harvester ant colony. The *Aphaenogaster* workers ran around in circles on the nest mound and within ten minutes went back to their own nest. The encounter with another species excited the harvester ant patrollers.

These perturbations were calibrated to mimic ordinary events in the life of an ant colony. Wind and flooding cause debris to appear on foraging trails and nest mounds, in amounts that have an effect like that of the barriers and toothpicks. Alien ants, such as the *Aphaenogaster cockerelli* used to stimulate patrolling activity, do trespass onto red harvester ant nest mounds.

The perturbations were repeated every day for three to six days (durations differed in different experiments). Every hour during the morning activity period (from about 6 A.M. until noon) I counted the numbers engaged in each of four tasks (foraging, nest maintenance, patrolling, and midden work) in a group of colonies. In each experiment, five to seven colonies were subject to a perturbation, and seven were undisturbed. I did the experiments over several days because days differ from each other. By using the same colonies day after day, I could distinguish day-to-day from colony-to-colony variation.

Taking the behavior of undisturbed colonies as a baseline state, I compared it to the behavior of colonies I interfered with. I asked, for example, how does the number of foragers in colonies with barriers on their foraging trails differ from the number of foragers in undisturbed colonies without any barriers? The results show that ant tasks are interdependent. The ants performing one task respond when something interferes with a different task. All of my interference experiments, each affecting one activity, changed the behavior of ants performing other tasks that I did not interfere with. For example, when numbers of patrollers were experimentally increased, numbers engaged in foraging also increased.

Foraging and nest maintenance are strongly related. When numbers of foragers were experimentally decreased by putting out barriers on the foraging trails, the numbers of nest maintenance workers increased. Even though the nest maintenance workers were not directly affected by the barriers on the foraging trails, some interaction between nest maintenance workers and foragers must regulate the numbers of nest maintenance workers. When numbers of nest maintenance workers were experimentally increased, by putting out toothpicks, numbers of foragers decreased. Again, though foragers were not directly affected by the presence of toothpicks, they responded to changes in numbers engaged in nest maintenance.

ANTS SWITCH TASKS

Why do numbers in one task affect numbers in another? One explanation is that ants switch from one task to another. If an ant leaves one task to perform another, numbers in the task it left will go down by one, and numbers in the task it went to will go up by one. Foraging and nest maintenance are linked in a way that seems easily explained by task switching: When my experiments caused foraging to diminish, nest maintenance intensified, and when my experiments caused nest maintenance to intensify, foraging diminished. Both results would be explained if foragers switched tasks to do nest maintenance. For example, when more ants were needed to do nest maintenance work, some of the foragers might have stopped foraging and switched to do nest maintenance work.

To find out if ants switch tasks, I did the perturbation experiments again with ants marked with tiny drops of model airplane paint. Before marking the ants we first cooled them down so that we could handle them without being stung, and so that they would not wave their antennae around and get them stuck in the wet paint. Making ants too cold to move was not easy in the desert, but eventually I found the ideal piece of equipment: an ice-cream maker that stays cold in the desert for a long time after being frozen. Paints of several colors were used, each color corresponding to one task. We collected ants as they foraged, patrolled, and did nest maintenance work or midden work, then marked them, let the paint on them dry, and released them. The next day we used some of the colonies with marked ants to do perturbation experiments and left others undisturbed as controls.

In colonies that were not subjected to any experiments, few ants switched tasks. Foragers marked one day tended to forage the next, and so on for patrollers, nest maintenance workers, and midden workers. From the one to four days ants were marked, to the next day when observations were made, conditions did not change much.

But when my experiments created a need for extra workers, the ants switched tasks. They switched in some directions but not in others. The general pattern is a flow of workers into foraging from all other tasks. The flow seems to originate with the nest maintenance workers, and once an ant leaves nest maintenance work, it will not go back.

To increase foraging activity, I put out piles of seeds. In response to a new, abundant food source, ants from the other three worker groups—nest maintenance, midden work, and patrolling—all switched tasks to forage. When cylinders and alien ants created a need for more patrollers, midden workers and nest maintenance workers switched tasks to patrol. But when toothpicks appeared, and more nest maintenance workers were needed, the newly recruited nest maintenance workers that came out to clear the toothpicks were unmarked ants, not ants that the day before had performed some other exterior task. The extra nest maintenance workers must have been recruited from ants that until then had worked only inside the nest. Ants that had already foraged, patrolled, or done midden work did not switch to help out with nest maintenance.

Nest maintenance workers seem to be a source of new workers in the exterior workforce. In many social insect species, younger workers first do tasks inside the nest, such as food storage and brood care, and older workers do exterior tasks such as foraging. Ants performing nest maintenance are in transition from inside to outside, because nest maintenance work is done partly inside the nest, as ants pile up sand from excavated tunnels, and partly outside the nest, as sand is carried out of the nest entrance. Once a young nest maintenance worker switches to a fully exterior task such as patrolling or foraging, she no longer reverts to tasks that take her back down to deeper regions of the nest.

The ways that harvester ants switch tasks make sense for a species competing for food. In the dynamics of task switching, foraging has priority. The system channels workers into foraging when a new food source appears.

TASK SWITCHING IN RESERVES

Only 25 percent of a mature red harvester ant colony works outside the nest. In many ant species, large numbers of workers usually do nothing, but will become active in unusual situations. I wondered whether there are reserve workers like this in a red harvester ant colony, and whether the reserve workers are themselves likely to switch tasks. I repeated the perturbation experiments with marked reserve workers. On day 1, I marked all the workers I could that were engaged in what I shall call task A. On day 2, I did a perturbation that increased the numbers of ants engaged in task A. I then used a different color to mark the workers that came out to do task A on day 2 but hadn't been marked the day before. These workers, who responded to a sudden need for extra help with task A, are the reserve task A workers. Then, on day 3, I did a perturbation that increased the numbers of workers engaged in another task, task B. The question was whether the marked reserve task A workers would switch to help with task B.

The directions of task switching by reserve workers resemble those of other exterior workers. Ants originally recruited as reserves to help with nest maintenance, midden work, or patrolling will switch to foraging when extra foragers are needed. Ants recruited to do extra nest maintenance work come from a pool of ants that have previously not worked outside the nest.

EACH ANT DECIDES WHETHER TO GO OUT AND WORK

Ants switch from one task to another, and this versatility explains why the numbers in different tasks are related. But task switching is not the whole story. There must be other ways that ants of different task groups communicate. The relation of numbers performing nest maintenance and foraging suggests task switching: When events drive nest maintenance up, foraging goes down, and when events drive foraging down, nest maintenance goes up. A tidy explanation for this would be that foragers switch tasks to help with the extra nest maintenance work. But foragers never switch back to nest maintenance. This means that foragers and nest maintenance workers must communicate somehow. When ants are recruited from inside the nest to perform extra nest maintenance work, the foragers stay inside the nest, inactive.

Why, when there is more nest maintenance work to do, do foragers tend to remain inactive, inside the nest? There is no mechanical interference; by the time the foragers emerge, the toothpicks have been moved to the edge of the nest mound where passing ants walk right over them. The puzzling inactivity of foragers on the day of a nest-maintenance-increasing experiment mimics a long-term oscillation between nest maintenance and foraging, in most colonies, over the course of a summer. It may be that my perturbation experiments trigger this cycle. If, because of an experiment, foraging decreases, this may be read by the colony, in some way, as the beginning of a low-foraging, high-nest-maintenance period.

The perturbation experiments showed that task forces interact. Ants do respond to events that directly affect only the ants in another task group. Marking individuals showed that there are two ways that ants adjust their behavior. From one hour to the next, an ant may switch from one task to another, and it may change its active/inactive status, becoming inactive when it was active before, or the reverse. These changes at the individual ant level adjust the task allocation of the colony, altering the numbers of workers engaged in the colony's tasks.

WORK IN OLDER COLONIES

As a colony grows older and larger, task allocation changes. The interactions of task groups have a different outcome in older and younger colonies. This is the most intriguing result of the perturbation experiments, and the one most difficult to explain. Ants live only a year, so an older colony does not contain older ants.

Colonies differed according to age in their responses to the perturbations. I performed the experiments with two classes of colonies: two-year-olds, and colonies five years or older. For older colonies, the worse things got, the more they compensated. More interference led to behavior more like that of undisturbed, control colonies. For younger colonies, however, more interference meant a stronger response, more different from undisturbed control colonies.

To increase the magnitude of interference, I imposed two or more perturbations at a time. Some colonies received toothpicks as a single perturbation, some received barriers as a single perturbation, and so on. Another group of colonies received both toothpicks and barriers at the same time. I did all three single perturbations of patrolling, foraging and nest maintenance in various combinations. These were all done to younger and older colonies at the same time. I went around all the colonies, including undisturbed ones for comparison, once an hour, and recorded the numbers of ants engaged in foraging, nest maintenance, patrolling, and midden work. If at the end of an hour's round there were any minutes left in the hour I sat down on a little stool halfway between the last colony and the first one of the next hour, and read a few paragraphs of *War and Peace* (which I also read in longer stints at night). After a few days

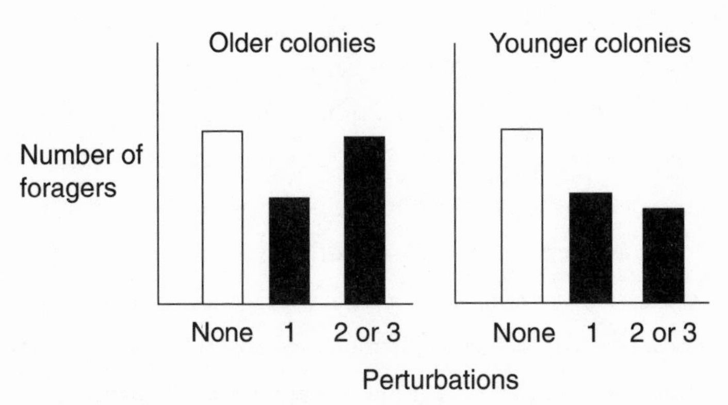

Older and younger colonies respond differently to perturbations. In younger colonies, the result of two perturbations together is close to the sum of each one when performed separately. In older colonies, the result of combined perturbations is not additive; the result resembles an undisturbed colony.

of this regimen I began to feel that even colonies burdened with toothpicks, barriers, and cardboard cylinders all at once had a simpler life than I.

Later when I looked at the data a surprising result emerged. The more an older colony was disturbed, the more it acted like an undisturbed colony. The response to two or three perturbations, in combination, was less than the sum of the responses to each one singly. For example, putting out toothpicks alone led to an increase in nest maintenance work and a decrease in foraging. Putting out barriers alone also decreases numbers foraging. Thus each single perturbation leads to a decrease in foraging, relative to the numbers foraging in undisturbed colonies. The response to the two experiments together, both toothpicks and barriers at once, should have been a more severe decrease in numbers foraging. More precisely, the additive response would be a decrease in numbers foraging (relative to numbers foraging in control colonies at the same time) equal to the sum of the decreases in foraging when each of the two perturbations were performed singly. But this is not what happened. Instead, numbers foraging were more like those in undisturbed colonies. Older colonies decreased foraging less in response to the two perturbations simultaneously than they did in response to either one alone. It seems that as disturbance increases in magnitude or complexity, an older colony compensates, sacrificing foraging effort less and less (see illustration on page 131).

The younger, smaller colonies reacted more to combined perturbations than to single ones. Foraging decreased more in response to combined perturbations than it did in response to the single ones.

Older colonies were more consistent, from week to week, than younger ones. I performed the same experiments, week after week, with several different groups of older colonies and with several different groups of younger colonies. Week after week, each group of older colonies responded much as the other groups did, but week after week, each group of younger colonies had a different response. Curiously, in any given week, the variation among younger colonies was no greater than the variation among older colonies. The differences were only in week-to-week comparisons. Apparently younger colonies are more susceptible than older ones to changes in weather or to the amount of food available. In one week, a given perturbation pushes the young colonies in one direction; in another week, it pushes them in another.

Why does task allocation depend on colony age? An ant lives only a year, so age differences in colony behavior cannot be explained by differences in the ages of individual ants. An older colony may appear to have a wiser foraging strategy; as environmental conditions become more disturbed, it devotes more effort to foraging. Disturbances might indicate worse to come, and it might be prudent in such conditions to stock up on food. But this apparent wisdom is not conferred by the advice of older, more experienced ants. Instead, something about colony organization must change as the colony grows older and larger.

The queen is the only ant that persists throughout the life of the colony, but the queen, despite her imposing title, merely lays eggs and does not direct the workers in any way. It is possible that the ants of each year somehow inherit information from previous cohorts of their sisters. A simpler, and thus more likely, explanation is that age differences in task allocation are a consequence of colony size. The process that determines which ant does which task, and whether an ant pursues a task actively, might be the same in older and younger colonies. But the outcome of this process might depend on the numbers of ants included in the system. Chapter 8 presents some models that provide examples of how this system could work.

CASTE

Interactions among task groups, which seem so obvious in human societies, have been little studied in social insects. This omission may be because the study of task allocation is still quite new, and it began with a research program bound up with the notion of "caste." "Caste" was originally used to refer to status in human societies in which people are thought to inherit particular social roles along with propensities for particular types of work. For social insects, "caste" referred originally to the distinction between reproductive individuals, such as ant queens, and sterile, nonreproductive individuals, such as ant workers. Whether an individual social insect will reproduce usually is determined by the time the adult emerges from the pupal stage. Ant queens tend to be much larger than workers, and have wings that they use to fly to a mating aggregation.

Gradually "caste" in social insects came to refer, not to distinctions between queens and workers, but to distinctions among workers in the task that they perform. In a minority of ant genera (60 of 263), there are species with ant workers of a range of sizes. Earlier work postulated that an ant would do a single task throughout its life, and that this task would be the one most efficiently performed by an ant of that size. An analogous system of division of labor might occur even in species where all workers are the same size. Ants might specialize throughout their lives in a particular task. In this sense, then, "caste" would not refer to any physical characteristic, but to an inherent tendency for an individual to do a particular task.

Why did this view of colony organization inhibit the study of interactions among task groups? I think that the notion of task as innate reinforces a view of individuals as independent of each other. The problem posed by the caste-oriented research program was to explain why colonies of a given species contain a particular distribution of workers in each caste. That is, colonies were supposed to contain a certain number of ants specialized to forage, another number specialized to do nest work, and so on. The problem was to explain how natural selection acts to produce optimal distributions, called "adaptive caste distributions." An assumption of this research program was that the number of workers that perform a task is a simple function of the number of specialized individuals in the corresponding caste. For example, the amount of foraging a colony does would correspond to the number of foragers it contains.

But the number of ant workers that perform a task is not a simple function of the number of ants of that task group. Ants switch tasks. An ant may specialize for a short time, but it often does several tasks, not just one. The number that perform a task right now depends on the rules that determine which ants will switch to perform that task in the current situation. Ants move from active to inactive and back. How much a colony forages may depend in part on how many ants are available to work as foragers, but it also depends on the numbers engaged in nest maintenance work, the numbers patrolling, and so on. Somehow, the behavior of ants in other task groups influences the probability that an ant will be active and the probability that it will perform a particular task.

The urge to locate the causes of behavior within individuals persists. One simple way to explain behavior is to say that an individual's task is genetically determined. Under close scrutiny, this explanation is inadequate. First, of course, we cannot explain how genes could determine behavior as complex as a social insect's task. Second, we know that ants switch tasks. Third, short-term changes must explain why an ant tends to perform a task actively in some conditions, but in other conditions remains inactive. These difficulties have led to more complex ideas about the ways that internal characteristics of individuals might contribute to task allocation.

Individuals might differ in their propensity to perform a task. In this view, a colony consists of individuals that have a range of thresholds, and each one will perform its task only when some external cue pushes it over the threshold. We could then ask, what is the system of external cues to which these individuals respond in varying ways? Or we could ask, what is the source of internal differences? Both questions are important. To investigate the system of external cues, it is reasonable to begin by assuming all individuals are alike, and then, once the external cues are understood, to investigate why individuals differ in response. If we try to figure out how individuals differ before we know what they are responding to, it is all too easy to confound individual differences with changes in external cues. When two workers act differently, is it because they are different or because they are responding to different things? The models described in chapter 8 take as a starting point the assumption that all individuals are alike. Eventually we hope to know enough to modify this assumption and investigate the ways that ants differ.

WHAT KIND OF PROCESS IS TASK ALLOCATION?

Though colonies must respond to changing conditions, the response does not have to be perfect. It is not like clock-work, or an army, each unit snapping into place so the whole system ticks on without a hitch. There must be enough ants to collect food, often enough for the colony to survive and grow. The appropriate numbers allocated to each task is really the appropriate range of numbers allocated over a set of similar occasions. If the colony did not get enough food today, perhaps it will tomorrow. The process results in more or less the right number of ants engaged in the appropriate task, often enough for the colony to carry on.

Maximizing numbers of ants that perform each task may not always be best for the colony. A task allocation problem for a human city is how to get the right number of firefighters to the scene of a fire. It may be a waste to have too many firefighters on the city payroll. Too many ants allocated to each task may be expensive for a colony if the excess ants could be doing something more useful than waiting around when they are not needed.

The most difficult thing to grasp about task allocation is that it is not a deterministic process even at the individual level. An ant does not respond the same way every time to the same stimulus; nor do colonies. Some events influence the probabilities that certain ants will perform certain tasks, and this regularity leads to predictable tendencies rather than perfectly deterministic outcomes. The ant is jostled along in a stream of events that send it sometimes into one task, sometimes another. Task allocation is not a system in which each ant awaits the crucial event that defines its status forever. Like a twig in a turbulent river, an ant may tend to go in one direction but there are many places it could get washed ashore, to be picked up and then swept in another direction altogether.

8

COMPLEX SYSTEMS

•

I saw them hurrying from either side
and each shade kissed another, without pausing,
Each by the briefest society satisfied.

(Ants, in their dark ranks, meet exactly so,
rubbing each other's noses, to ask perhaps
What luck they've had, or which way they should go.)

Dante, *Purgatorio,* Canto XXVI

•

The dynamics of ant colony life has some features in common with many other complex systems: Fairly simple units generate complicated global behavior. If we knew how an ant colony works, we might understand more about how all such systems work, from brains to ecosystems. Because we don't yet comprehend any natural complex system, I think it is premature to say how general a theory we may eventually achieve. The intriguing question about task allocation is how might an ant react to local events, in a simple way, that in the aggregate produces colony behavior? The same kinds of questions come up, over and over, throughout biology: How do neurons respond to each other in a way that produces thoughts? How do cells respond to each other in a way that produces the distinct tissues of a growing embryo? How do species interact to produce predictable changes, over time, in ecological communities? These are the big, general questions of biology, and many of us dream that when we have the answers, from different fields of biology, it will be possible to see similar processes at work from cells to ecosystems.

We know enough about task allocation in harvester ants to ask very specific questions. How do colonies adjust their behavior to meet the contingencies of a changing environment? Why do older, larger colonies behave differently from younger, smaller ones? One way to approach such questions is to make mathematical models that mimic the behavior of colonies. Our hope is that such mimicry will be a reliable guide to empirical work. We set up a mathematical system and look at how it behaves. If data on real ants match the model's predictions, the model may mimic the way real ants operate. But it might not. A computer simulation may have ants writing Shakespeare's sonnets, but we should not expect any poetry from ants until we know whether real ants use algorithms like the ones in the computer program. Models offer possible ways that an ant colony might work, to guide our investigation of real ants. Model-making goes along with empirical work that investigates whether the kinds of behavior suggested by the model take place in real ants. This line of work, in turn, can stimulate the creation of better, more informative models.

In a very general way, it is clear what kind of models we need, because one basic feature of the organization of social insect colonies is already obvious. Individuals, following simple, local rules, generate the achievements of colonies.

One class of models describing how simple, local rules generate global complexity are the neural networks. We used a neural network or parallel distributed process model to describe interactions among worker groups in a harvester ant colony. It draws on the analogy between colonies and brains. In both systems, relatively simple units (ants or neurons), using local cues, can achieve complex, global behavior. In such models, the individual units are usually called "neurons." We imagined each unit as an ant. The only information available to each individual unit is from its interactions with other units, interactions so simple that they can be designated as either positive or negative, like an on/off switch or a neuron that either fires or does not fire. Globally, such systems can generate complex patterns. In a neural network model, whether an individual is on or in the "+" state depends on a weighted sum of its interactions with other individuals. If the sum of interactions is greater than some threshold number, the individual enters the *on* state; if the sum is below the threshold, the individual enters the *off* state. The *on* and *off* states of the units are assigned an energy value, and the accumulation of many of these interactions eventually settles in to an equilibrium.

I worked on a neural network model with Brian Goodwin, a mathematical biologist, and Lynn Trainor, a physicist. At the time, Brian Goodwin was a professor at the Open University in England and Lynn Trainor was a professor at the University of Toronto. The model simulates the allocation of workers to various colony tasks, such as foraging, nest maintenance work, and patrolling. We simulated a neural net in which each unit, or ant, could be in one of eight possible categories. In the model, each ant could be either active or inactive, and it could belong to one of four possible task groups: midden workers, patrollers, nest maintenance workers, or foragers. Each ant's decision about activity and about task depends on the sum of pairwise encounters with other ants. The results of the computer simulations had this in common with the results of perturbation experiments: an initial increase or decrease in numbers of ants in one category propagated to other tasks, changing the numbers in other categories as well.

The neural network model shows that the relation I observed among task groups could result from accumulated interactions between pairs of ants. If the organization of ant colonies corresponds to the system described by this model, what must real ants do? Each ant must be capable of perceiving the task group and activity status of the other ants it encounters. Combining this information according to some simple rule, such as the sum of interactions, each ant could decide whether to continue the task it is performing, switch tasks, or remain inactive inside the nest.

The neural network model describes only one of the ways in which simple, local rules could generate the allocation of workers into various tasks, so that events changing numbers engaged in one task will change the numbers engaged in other tasks.

Next I collaborated with Steve Pacala, an ecologist at Princeton, and Charles Godfray, an ecologist at Imperial College at Silwood Park, part of the University of London. This collaboration began when I was working at the Centre for Population Biology at Silwood Park, Steve Pacala was visiting there, and Charles Godfray was, as he is still, down the hall in the Department of Biology. We worked on a second model of task allocation. In this deterministic model, an individual's decision is based on two kinds of stimuli: the rate of interaction with other workers, and the state of the environment relevant to a particular task. The combination of interaction rate and environmental stimuli makes this model more realistic than the neural-network one, in which individual decisions are based only on interaction with others. There is overwhelming evidence that task allocation is affected by cues from the environment. Otherwise, colonies could not respond to changes in environmental conditions by reallocating workers to various tasks—but of course they often do so.

The deterministic model predicts the numbers of ants allocated to each of Q tasks. These numbers change, as a result of interactions between ants and how successful they are at their task. An ant is inspired to perform a task when it meets another ant successfully engaged in that task, and it gives up a task when it is unsuccessful at that task.

The model works as follows. An ant is either actively engaged in one of the Q tasks, or else it is inactive. If it is actively engaged in one of the tasks, it is either successful or unsuccessful. Success depends on current environmental conditions; for instance, few foragers will be successful when there is little food available. Unsuccessful individuals engaged in task *t* will quit task *t* and become inactive. The time it takes for an ant to quit depends in part on the environment, which determines how likely the ant is to be successful or unsuccessful. The model assumes a unique quitting rate for each task. Differences among tasks in the quitting rate might depend on the cost of performing a task. A high-cost task might not be worth pursuing when workers engaged in it cannot succeed. For example, if foragers are unsuccessful at collecting seeds, they may be losing more water than they gain. At this point, they might quit foraging and remain inactive inside the nest until more food becomes available. In contrast, ants working inside the nest to enlarge a chamber may not be using as much energy. Suppose they hit a rocky patch of soil in which few individuals can succeed in scraping away at the wall of the chamber. These unsuccessful nest workers may continue longer before they quit than would unsuccessful foragers.

Whether an ant performs a task depends on how successful it is, and if it is not successful, on the rate at which ants attempting that task just give up. Both of these factors are related to the environment. In the model, a second kind of influence is the interactions an ant experiences. When an unsuccessful ant meets a successful ant from a different task, it becomes likely to switch to the new task of the successful ant.

In the model, the numbers performing a task *t* are determined by:

(1) subtracting the number of individuals that are unsuccessful at task *t* and quit, thereby becoming inactive, (2) adding the number of individuals from other tasks that encounter successful task-*t* ants and switch tasks to perform task *t*, (3) subtracting the number of individuals that are unsuccessful at task *t*, encounter successful individuals from another task, and switch from task *t* to the other task, and (4) adding the number of inactive individuals that encounter successful task-*t* individuals and switch to perform task *t*.

COLONY SIZE AND ABILITY TO TRACK A CHANGING ENVIRONMENT

An interesting result of this model is a prediction of how well task allocation can track a changing environment. Tracking speed is related to colony size. Large colonies should be able to track a changing environment more rapidly than small ones, because the larger the colony, the higher the rate of interaction. High interaction rate means that successful ants interact often with unsuccessful ones, frequently transferring information about the environment.

But there is a disadvantage to large colony size. In a large colony, a successful ant might interact with large numbers of unsuccessful or inactive ones that would then be recruited to the task of the successful one. As a consequence there might be more workers performing the task than the situation warrants. For example, suppose a successful forager comes back to a large colony and meets large numbers of unsuccessful or inactive workers inside the nest. This encounter would lead many workers to switch to forage, possibly more than would be warranted by the amount of food available. If interaction rates are too high, too many workers might be active in unprofitable environments.

This suggests an explanation for why some ants regulate contact at high density. As noted in chapter 6, we found that ants of *Lasius fuliginosus* do so. Perhaps in this species, encounter patterns influence an ant's task as they do in the model. If encounters with successful ants encourage others to switch to that task, too much interaction might lead the colony to allocate more ants to a task than the environment really warrants. Little is known about task allocation in *L. fuliginosus.* This species lives in beautiful places, nesting at the base of old oak trees in the forests of England and northern Europe. More research is needed to find out why these ants regulate contact rate, and such research might be very pleasant.

THEORETICAL WORK NEEDED

The models of task allocation we now have, including the two described above, are not yet ready to work with. They produce results that resemble some observations from real colonies, but this outcome does not demonstrate that the models describe the way the ants operate. The next step is to test whether real ants behave in the way described by any model. This is not yet possible, because the predictions of the models are too general to match up with data. Still needed are models that predict numbers of ants engaged in particular tasks, in a form that can be compared to counts of numbers of ants engaged in particular tasks in real colonies. We have to include more of the details of what ants actually do, such as spatial behavior and temporal patterns of activity. The difficult part, as in any modeling exercise, is to figure out how many of the details are needed to give realistic predictions. Another complementary approach is to zoom in on one or two tasks, or one time of day, to shrink the window of empirical detail that must be added to the model. As Richard Levins wrote, models tend to be so general that they cannot make predictions about particular systems, or so detailed that they merely rephrase what is already known about a system. The challenge is to resist the gravitational pull of these two extremes and find the intermediate position, useful general predictions that can be evaluated by actual data.

INTERACTION PATTERNS

The two models described above have something in common. In each, an ant's behavior is influenced by the pattern of interactions it experiences. Many other models of colony organization share this feature. For example, Tom Seeley and his colleagues simulated the allocation of honeybee foragers to different food sources, using a model in which an individual bee responds to the rate at which it encounters other bees.

Jean-Paul Deneubourg, Simon Goss, and their colleagues have used a different type of model, similar to the deterministic one described above, that shows the workings of initially random processes that have an irreversible, in some ways unpredictable outcome. A foraging trail can be seen as an irreversible outcome of the behavior of many randomly searching ants. The destination of the trail is not completely predictable if there are multiple food sources. Once a trail forms, and most searchers join the trail, the location of the trail has been irreversibly determined. In their models, each individual's behavior is represented as simple rules that involve random searching and interaction, where the interaction is following the pheromone trail of another individual. Such rules can produce branching foraging trails.

Models based on interaction patterns are especially appealing because task allocation in harvester ants is related to colony age and size. The changes in task allocation as colonies grow older and larger are consistent with a process that involves interaction patterns. Interaction rates should depend on colony size; rates should be higher when there are more ants to interact with and lower when the numbers or density of ants is low.

The models show how task allocation could work if an ant's behavior depends on simple cues from its environment and from its interactions with other ants. Whether harvester ants actually use interaction patterns in task allocation is an empirical question.

INTERACTION HISTORY AND TASK ALLOCATION

I began thinking about interaction patterns in harvester ants because of the way the patrollers acted when we collected them to mark them with paint. To study task switching, we collected ants while they were performing some task, marked them with paint, released them, and then looked at which task they performed the next day.

We collected ants with an aspirator, a tiny vacuum device that very quickly lifts the ants into a tube and deposits them in a vial. Collecting foragers was easy; foragers didn't seem to react at all when their fellow foragers suddenly disappeared from the trail. We were just another predator, like the horned lizard that stands beside the trail sucking up ants while the surviving foragers keep walking by, apparently heedless. Collecting midden workers was easy too. Collecting nest maintenance workers required a bit more care, because if the person with the aspirator breathed out instead of in, the gust of strange smells into the nest entrance sometimes sent the nest maintenance workers running back to cower in the entrance, with their heads sticking out and their antennae waving around in perplexity.

Collecting patrollers was completely different. Even the most careful aspirating of only a few patrollers, well apart from each other, could cause the whole colony to shut down for the day: The nest maintenance workers and other patrollers would go back into the nest, and later the foragers would not come out at all. It seemed that foraging could not commence unless returning patrollers somehow signaled that the outside world was safe at the time. This led me to think that foragers, waiting inside the nest entrance, respond to the rate or number of returning patrollers; when too many disappeared, and the rate or number fell below some threshold, the foragers stayed inside.

What puzzled me most about this observation was the rapidity of the patrollers' reaction. When some patrollers outside the nest disappeared, the rest of the patrollers sometimes headed back into the nest immediately, within seconds—long before there was time for anyone to go back into the nest and assess the rate at which patrollers were returning. This happened even when we put the aspirator right over the patrollers and never exhaled through it, so that there would be little opportunity for a cloud of alarm pheromone to escape. The patrollers still reacted, often from a distance that seemed too great for any pheromone to travel so quickly.

Maybe the patrollers expect to meet each other outside the nest at a certain rate. We found that the colony was most likely to shut down if we took patrollers in the initial phase of patrolling, when they circulate around the nest mound, often stopping to touch antennae with each other. They look like solitary strollers in a large park, all of whom have lost their dogs: Walk a little way, see someone, ask, "Have you seen a large white dog? No? Thanks," then move on a short way until the next encounter. Later in the morning, when the patrollers move off the mound to search the surrounding area, and meet each other much less often, we could sometimes manage to take patrollers with no effect on later foraging. Perhaps on the nest, where the patrollers expect to meet very frequently, removing a few decreases the encounter rate so drastically that the patrollers react immediately. Once the patrollers are off the foraging mound, if the usual rate of interaction is much lower, the disappearance of a few patrollers might not change each ant's rate of encounter much, or not change it very quickly.

This led me to wonder what determines how often the ants meet and how much the encounter rate will change if a few ants disappear. I started to think about how patrollers move. If every patroller left a colored line in its path, what pattern of patroller paths would cover the nest mound? It seems obvious that the shape of the paths would determine how often they meet. If they all left the nest mound and took off in straight lines, they would never meet. If they each went to a different place on the mound and then turned around and around in tiny circles, they would never meet. But somewhere in between these two path shapes, patrollers would meet each other, and how often must depend on the way that they move. How exactly encounter rates depend on path shape proves, however, to be a very complicated question.

I was excited about the possibility that I was seeing patrollers react to a change of interaction rate—though their reaction meant it took us two summers' hard work to mark enough ants to find out about task switching in patrollers—because of the link between interaction rates and colony size. If the behavior of ants is influenced by their recent interaction history, then colony size should affect ant behavior, because interaction rates should depend on colony size. This could explain why task allocation differs between young, small and old, large colonies.

WHAT EXACTLY IS AN INTERACTION?

Interactions between ants can take many different forms. The form best studied so far is the transfer of a chemical cue. A worker exposed to alarm pheromone may stop performing its task and begin to circle around and perhaps to attack. The worker is responding to an interaction, in the form of transfer of an airborne chemical cue, with another worker. In the same sense, people who rush out of a building when someone shouts "Fire!" are responding to an interaction with the shouter. By analogy, two neurons interact by transferring chemicals across the gap between each one's extending branch.

Interactions can occur on many timescales. Ants interact with a time delay when one ant responds to a chemical cue deposited earlier by another one. An example is trail pheromone; many minutes may elapse between the time that a successful forager, returning to the nest, puts the scent down, and the time that a newly recruited forager finds it on the way out to the food.

I have concentrated so far on one form of interaction: brief antennal contact. Anyone who has ever watched ants has seen two ants approach each other, briefly touch antennae, then go on their separate ways. The antennae are the organs of chemical perception; ants smell with their antennae. In the course of a brief antennal contact, an ant can distinguish whether the other one is a nestmate.

An interaction may not transmit any message besides the incidence of interaction itself. An ant may respond to the number or rate of interactions. Advertising people speak of the number of "hits" a consumer receives, encounters with the name of a product. The idea is that whether you buy Coca-Cola may depend not on what you learn about the qualities of Coca-Cola, but just on how many times, or how often, you encounter the name "Coca-Cola." For ants as well as people, the interaction pattern may be more important than the message.

ENCOUNTER PATTERNS AND TASK ALLOCATION IN HARVESTER ANTS

We know, from the perturbation experiments described in chapter 7, that task groups are interdependent. Events that affect one task group affect another. How does an ant perceive an event that affects another task group? Either the ant perceives something about the task itself, or the extent to which it has been accomplished, or the ant perceives something about the ants performing the other task.

There are some indications that in harvester ants, numbers of workers, not the extent to which a task is accomplished, provide the cues that generate changes of worker allocation. First, it is easier to see how a worker could track numbers of workers than how it could appraise the status of various tasks. Workers engaged in different tasks tend to be spatially segregated (see illustration on page 37). For example, harvester ant foragers leave the nest entrance and go directly to a foraging trail that can extend 20 to 30 m. Nest maintenance workers often travel only a few centimeters from the nest entrance, where they put down debris collected inside the nest, and they rarely leave the nest mound. Debris near the nest entrance will be discovered by nest maintenance workers, but foragers will not encounter the debris unless it is in their path from the entrance to the foraging trail. It seems unlikely that when more nest maintenance work is needed, foragers will apprehend the necessity directly.

Although workers engaged in one task are unlikely to encounter cues that indicate the status of other tasks, they are very likely to encounter workers from other task groups. All exterior workers go in and out of the nest frequently. Colonies are active outside the nest for six to seven hours each day. One trip by a forager may take about half an hour, and nest maintenance workers, patrollers, and midden workers all go in and out of the nest even more often. Exterior workers rarely travel deep inside the nest. This means that workers of all exterior task groups, as they move in and out of the nest will mix in the upper chambers directly inside the nest entrance. Ants of different task groups may be spatially segregated in work outside the nest, but encounters among them are frequent as workers go in and out of the nest.

Ants working outside the nest make frequent trips back into the nest, and exterior workers seem to remain clustered in the upper chambers when they are inside the nest. Thus the numbers active outside the nest will influence the numbers of interactions experienced by all exterior workers, both the active ones when they enter and leave the nest, and the currently inactive ones that are simply waiting inside.

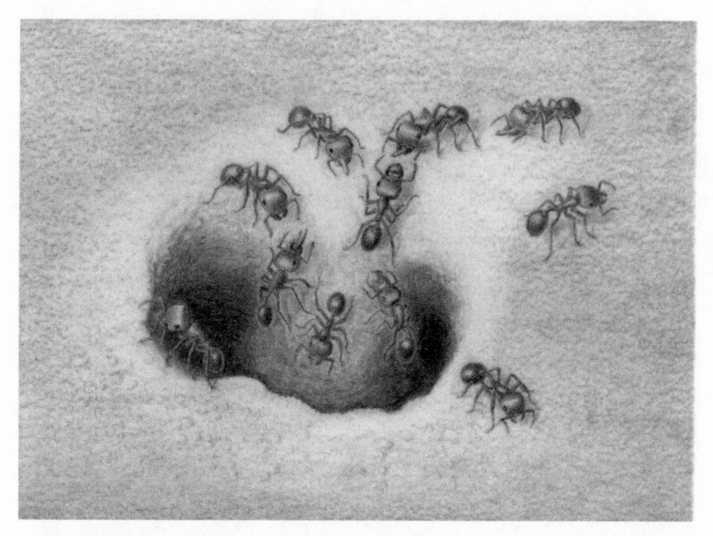

Ants engaged in different tasks meet and touch antennae as they come in and out of the nest.

Suppose the probability that an ant performs a task depends on its recent encounter history. The ant distinguishes encounters with ants of different task groups. Its probability of task performance depends on a rule such as "If I meet five successful foragers, then go out to forage; if I meet less than five foragers, then remain inactive." The decrease in numbers of ants leaving the nest to forage when I took the food away from returning foragers suggests that some such rule is at work: Ants that met too few successful foragers were less likely to leave the nest to forage.

I have tried to watch encounters inside harvester ant nests with a fiber-optics microscope. I thread the wire into the nest and look through the microscope in my hand. Though the ants react to the light at first, they soon calm down and rush past it. The lens is extremely wide angle, so the ants loom alarmingly into view, like monsters in a horror movie, then quickly disappear over a distant horizon. I tried using red light, which ants can't see, but the ants' behavior was so similar to that of ants in white light that I went back to the white light so I could see more. How much I can see depends on the shape of the tunnels in a particular nest. Usually the chamber just inside the entrance has two or more tunnels leading into it, and each of them branches quickly into more tunnels or chambers. I can rarely get the microscope down past two or three branching levels. Older nests tend to have smoother walls and floors. It is more difficult to follow ants in older colonies because the entrance leads to a more elaborate system of branches than in the nests of 1- or 2-year-old colonies.

The ants usually are not milling around in a great festival of antennal contact, although they do occasionally convene in this way. Instead some ants seem to be standing around doing nothing, and others rush past them, both downward and upward bound. Some ants seem to come up decided on their task, frequently carrying soil or refuse in their mandibles, and rush out immediately. Others come up, join the ants in the entrance chamber, and eventually leave the nest. Foragers run in, often nonchalantly drop their hard-won prizes on the floor and continue down into a deeper chamber. Another ant may come by, stumble upon the seed on the floor, and take it deeper into the nest. It is difficult to keep track of marked ants inside the nest because there are so many ants there and I cannot control the side from which I look at them. The limitations of the fiber-optics microscope, at least so far, have taken me back to the lab.

We keep laboratory colonies on a table (we find brood production is much higher if the table is vibration-free), in a series of plastic boxes connected by Tygon tubing to a wooden foraging arena. The foraging arena has several tiers, supported by pillars and connected by access ramps, so the whole thing resembles a crude wooden wedding cake. Tygon tubing leads from the entrance of the arena to one plastic box that we call the "outer chamber," which is connected from the other side to many rows of plastic boxes that serve as a nest. The ants of the outer chamber and foraging arena do not mix much with the ants in nest boxes. In real harvester ant nests, exterior workers rarely go down into the deeper chambers. The outer chamber in the lab seems to be used like the upper chambers of nests in the field.

To find out how an ant's recent interaction history influences its task, we mark it, follow it, and record all its antennal contacts with other ants. We analyze the data to see whether there is any association between whom an ant met and what that ant does next.

Laboratory colonies spend an inordinate amount of time on midden work. There are many more dead ants to deal with in the laboratory than in the field, because in the field other scavenging species tend to take the dead ants away. Thus, results so far are mostly about midden work.

An ant's task depends on its recent encounter history. The more contact an ant has with midden workers, the more likely it is to do midden work. When an ant apparently not doing much, walking or standing around, meets but few midden workers, it continues to walk around. The more midden workers an ant meets while walking around, not on the midden, the more time it is likely to spend on the midden. Rate, as well as number, of contacts is important. Frequent contact with midden workers increases the probability that an ant will switch to midden work.

For ants to respond to the number or rate of encounters, they must distinguish other workers according to the tasks they are performing. Ants performing a certain task may have a characteristic odor. By grooming themselves and each other, ants cover their exoskeletons with a greasy layer of fatty acids called cuticular hydrocarbons. The colony-specific odor that ants use to distinguish nestmates from non-nestmates is contained in the cuticular hydrocarbons. With Diane Wagner and others, we compared the cuticular hydrocarbons of different task groups in harvester ant colonies. We found that task groups differ in cuticular hydrocarbons, and the differences are similar despite variation among colonies. Foragers and patrollers show a greater abundance of n-alkanes than nest maintenance workers, and nest maintenance workers show a greater abundance of monomethyl alkanes, dimethyl alkanes, and alkenes, than patrollers and foragers.

Because ants switch tasks, it is unlikely that ants of each task group secrete different cuticular hydrocarbons, though it may be that the hydrocarbon profile of older ants, such as foragers, differs from that of the younger nest maintenance workers. Probably the task itself changes the hydrocarbon profile of an ant. Foragers spend much longer out of the nest, searching for seeds, than nest maintenance workers do in their brief trips to deposit excavated soil near the nest entrance. Perhaps UV light transforms the chemicals on an ant's body, so that after many foraging trips, an ant comes back smelling different from the way she did as a nest maintenance worker.

There is a great deal left to learn about how a harvester ant's environment affects its behavior. The process described above is one of positive feedback; more midden workers means more contact with midden workers, which will lead to still more midden workers. What prevents the colony from getting stuck with all ants performing midden work? There must be some negative feedback, probably from the environment. Midden workers in the red harvester ant, as well as in another species of the same genus, *P. badius,* may help prevent intrusion on the nest mound by other ant species, by transferring some scent-mark to midden material. Perhaps when the midden is not sufficiently permeated with the scent-marking chemical, a few ants may respond to the lack of midden scent and begin midden work, and their encounters with other ants will elicit more midden work. Then midden work might slow down if workers become more likely to return to the nest when they do not encounter midden material that requires scent-marking.

To understand fully how task allocation operates in harvester ants, we would need to know how interaction patterns combine with environmental cues to set the probabilities that an ant will perform each task. We know that the outcome of these dynamics is predictable: More patrollers means more foraging, more foragers means less nest maintenance, more nest maintenance means less foraging, and so on. We are not yet able to explain how these outcomes arise from the behavior of individuals.

How a harvester ant moves depends on its task (see illustration on page 37). The longer an ant's trips outside the nest, the fewer its opportunities for contact with ants of other task groups as they mix in the nest entrance. Foragers stay outside longer than midden workers, patrollers, or nest maintenance workers, and perhaps their long trips outdoors make them less susceptible to changing tasks, as they are unlikely to meet other workers who might distract them from foraging. Path shape will affect contact rate: The zigzag paths of patrollers may help to bring them into frequent contact with each other; the straight-line dumping missions of nest maintenance workers outside the nest means they rarely meet except inside the nest entrance.

Task groups may also differ in the way ants move inside the nest.

We will need to know more about task allocation before we can discover how it is evolving. We must understand the process well enough to find out if colonies vary slightly in task allocation, because variation is the starting point for natural selection. Suppose a forager in one colony needs 10 contacts with foragers before it will leave the nest to forage, but in another colony each forager needs only eight contacts to leave the nest. The latter colony might forage more. Small differences among colonies in the rules that influence how likely an ant is to forage, could lead to substantial differences in amount of foraging. When we can specify precisely the rules that influence task decisions, we can look for variation among colonies in those rules. Then when we know how task allocation varies, we can ask whether natural selection is shaping the way colonies work.

Why older colonies behave differently from younger, smaller ones remains an open question. If ants acted exactly like billiard balls, then the more ants per unit area, the more interactions. This would mean very different interaction rates in an old harvester ant colony, with 10,000 or more ants, than in a young one. Though ants do not act like billiard balls, how often ants meet depends on how they move around. But the details will be complicated. As the colony grows, the shape of the nest changes. Looking through the fiber-optics microscope I see more tunnels, and more branching in those tunnels, the older and larger the colony. The space in which ants congregate and meet becomes more differentiated as a colony grows. The chambers and tunnels near the nest entrance help to shape the flow and mingling of ants of different task groups. As the number of ants increases, the channels they live in change shape, and these changes modulate their encounter patterns. The development of task allocation in growing colonies is not a simple matter of increasing contact rates with increasing numbers of ants. To understand how task allocation depends on colony size, we will have to find new ways of modeling and observing the relations between how ants move around and how often they meet in crowded, curved, branching spaces.

EPILOGUE: LESSONS FROM THE ANTS

•

Many are the moral Instructions arising from the Sight of a Colony of Ants; with a few of which it may not be impertinent to close this Account. Their surprising incredible Affection towards the Young, might teach us to value Posterity and promote its Happiness. The Obedience they pay their respective Queens might read us a lecture of true Loyalty and Subjection. Their incessant Labours may serve to enliven the industrious, and enflame the lazy Part of Mankind. The unanimous Care exerted by each Colony for the common Emolument, might let us know the Consequence of Public Good, and tempt us to endeavour the Prosperity of our Countrymen. From their Economy we may learn Prudence, from their Sagacity Wisdom.

William Gould, *An Account of English Ants*, 1747

•

Emulating ants does not improve one's character. A person with the moral qualities of an ant would be terrifyingly empty. And I have not learned much about people from watching ants. People remind me of ants only when seen from so far away that they no longer resemble people; in the movie *Titanic*, the passengers scrambling up the sinking hull seemed to behave like ants.

But perhaps the ants have something general to teach us,

at least by analogy, about how nature works. Any system of units that lack identity or agency, whose behavior arises from the interactions of these components, has something in common with ant colonies. It may be that the same kinds of relations that link ants and colonies allow neurons to produce the behavior of brains, a host of different cells to produce immune responses, and a few dividing cells eventually to produce a developed embryo.

One lesson from the ants is that to understand a system like theirs, it is not sufficient to take the system apart. The behavior of each unit is not encapsulated inside that unit but comes from its connections with the rest of the system. To see how the components produce the response of the whole system, we have to track these connections in changing situations. You could dissect a brain into millions of separate nerve cells but would never find any dedicated to thinking about "nature," or "ants," or anything else; thoughts are made by the shifting pattern of interactions of neurons. Antibodies form in the immune system as a consequence of encounters with foreign cells. Ants are not born to do a certain task; an ant's function changes along with the conditions it encounters, including the activities of other ants.

All of the natural systems that work like ant colonies may have particular processes in common. But the more we learn, the more likely we are to discover differences. Understanding any natural system means learning about it in detail, and those details tend to make each system unique. We speak of "model systems"; the evolution of body shape in fruit flies as a model for the evolution of all animals; the reaction of mice to a drug as a model for the reaction of humans. In fact one natural system cannot represent another. Instead, a few examples provide a rough idea of how things work, but then counterexamples appear that whittle away at this idea, until eventually narrower, more specific, but more accurate ideas emerge.

Nevertheless it is tempting to speculate about the generality of interaction patterns as a source of information in natural systems. What I like about the idea that an ant's task decision is based on its interaction rate is that the pattern of interaction, not a signal in the interaction itself, produces the effect. Ants do not tell each other what to do by transferring messages. The signal is not in the contact, or in the chemical information exchanged in the contact. The signal is in the pattern of contact. Such a process might operate in brains, immune systems, or any place where the rate of flow of a certain type of unit, or the activity level of a certain type of unit, is related to the need for a change in the rate of flow. Interaction rate is the local translation of a characteristic of the whole system, rate of flow or activity, and each unit's reaction to this local cue contributes to a predictable response by the whole system.

The other intriguing feature of interaction patterns is that such patterns depend on the size of the system. The more units available to interact, the greater the rate of interaction each unit can experience. Like ant colonies, many living systems change size as they mature. If the parts of a system adjust their behavior according to the rate at which they interact with others, then the system's behavior can change as it grows even though the parts themselves continue to work in the same way. Ant colonies may not be the only natural systems to use this kind of process. But we need to know more about how ant colonies use interaction patterns before we can say whether such patterns are used elsewhere in similar ways.

Ants do not offer moral instruction, but they show how simple parts make complex living systems, and how those systems connect to the outside world. A colony's standing in its population depends on how many foragers it sends out, which hinges on the behavior of ants as they interact within a colony. Looking at ants in colonies while looking at colonies in populations, we see how the layers of a natural system fit together.

NOTES

•

An up-to-date list of publications is posted on our lab Web site: http://ant.stanford.edu/welcome.html

Notes refer to topics discussed on the indicated pages.

Chapter 1

Page 9 on nest-plugging by *Aphaenogaster cockerelli* (formerly called *Novomessor cockerelli*): 1988. Gordon, D. M. Nest-plugging: interference competition in desert harvester ants (*Pogonomyrmex barbatus* and *Novomessor cockerelli*). *Oecologia* 75: 114–17.

Chapter 2

Page 14 on estimating the lifespan of a red harvester ant queen: 1991. Gordon, D. M. Behavioral flexibility and the foraging ecology of seed-eating ants. *American Naturalist* 138: 379–411.

Pages 18–19 on mortality rate for founding queens: 1996. Gordon, D. M., and A. W. Kulig. Founding, foraging and fighting: colony size and the spatial distribution of harvester ant nests. *Ecology* 77: 2393–2409.

Page 19 The chart on page 19 comes from data described in: 1992. Gordon, D. M. How colony growth affects forager intrusion in neighboring harvester ant colonies. *Behavioral Ecology and Sociobiology* 31: 417–27.

Page 20 maps of the site from 1988 to 1993, showing that the colonies have become more crowded, are in the article by Gordon & Kulig, cited above, Founding, foraging and fighting.

Pages 24–25 on nest relocation: 1992. Gordon, D. M. Nest relocation in the harvester ant, *Pogonomyrmex barbatus. Annals Entomol. Soc. Am.* 85(1): 44–47.

Page 28 Counts from excavations of nests of 3 *Pogonomyrmex* species: 1981. MacKay, W. P. A comparison of the nest phenologies of three species of *Pogonomyrmex* harvester ants (Hymenoptera:Formicidae). *Psyche* 88: 25–74.

Page 30 MacKay reports that marked exterior workers are not found deep in the nest in the article cited for page 28, above.

Page 30 on the movement of ants around laboratory colonies. Pinter N., A. Fullerton, and D. M. Gordon. Temporal polyethism in the red harvester ant, *Pogonomyrmex barbatus.* Manuscript in preparation.

Pages 32–39 on the daily round of harvester ant colonies: 1984. Gordon, D. M. Species-specific patterns in the social activities of harvester ant colonies. *Insectes Sociaux* 31(1): 74–86.

1983. Gordon, D. M. Daily rhythms in social activities of the harvester ant, *Pogonomyrmex badius. Psyche* 90(4): 413–23.

Page 34 Patrollers choose the day's trails. See the first article cited above for page 14 and: 1983. Gordon, D. M. The relation of recruitment rate to activity rhythms in the harvester ant, *Pogonomyrmex barbatus. J. Kansas Entomological Society* 56(3): 277–85.

Page 35 On the function of the midden: 1984. Gordon, D. M. Harvester ant middens: refuse or boundary? *Ecological Entomology* 9: 403–12.

On the duration of foraging trips, see the article cited above in note for pages 18–19.

Chapter 3

Page 42 For a graph showing variation among colonies and day-to-day variation in foraging intensity, see: 1991. Gordon, D. M. Behavioral flexibility and the foraging ecology of seed-eating ants. *American Naturalist* 138: 379–411.

Pages 43–45 Foraging maps are described in detail in: 1995. Gordon, D. M. The development of an ant colony's foraging range. *Animal Behaviour* 49: 649–59.

Page 48 Patrollers choose the day's trails. See citations for page 34, Chapter 2.

Pages 51–55 On the distribution in space and time of seeds used by red harvester ants: 1993. Gordon, D. M. The spatial scale of seed collection by harvester ants. *Oecologia* 95: 479–87.

Pages 57–58 Some of these cartoons are shown in the article cited above for page 42.

Pages 61–65 experiment with enclosed colonies: 1992. Gordon, D. M. How colony growth affects forager intrusion in neighboring harvester ant colonies. *Behavioral Ecology and Sociobiology* 31: 417–27.

Pages 66–67 monitoring encounters between neighbors in undisturbed colonies: 1996. Gordon, D. M., and A. W. Kulig. Founding, foraging and fighting: colony size and the spatial distribution of harvester ant nests. *Ecology* 77: 2393–2409.

Pages 68–70 on neighbor recognition: 1989. Gordon, D. M. Ants distinguish neighbours from strangers. *Oecologia* 81:198–200.

Page 71 Foragers switch trails: See article cited for page 42.

Pages 71–73 on whether some foragers specialize in interactions with neighbors: 1997. Brown, M. J. F., and D. M. Gordon. Individual specialisation and encounters between harvester ant colonies. *Behaviour* 134: 849–66.

Chapter 4

Pages 75–77 how neighbors affect the success of founding colonies: 1996. Gordon, D. M., and A. W. Kulig. Founding, foraging and fighting: colony size and the spatial distribution of harvester ant nests. *Ecology* 77: 2393–2409.

Page 77 competition with neighbors does not kill established colonies: 1998. Gordon, D. M., and A. W. Kulig. The effect of neighboring colonies on mortality in harvester ants. *Journal of Animal Ecology* 67: 141–48.

Pages 78–80 measuring numbers of alates: 1992. Our alate trap was based on the trap used by Jim Munger in J. C. Munger. Reproductive potential of colonies of desert harvester ants (*Pogonomyrmex desertorum*): effects of predation and food. *Oecologia* 90: 276–82.

1999. Wagner, D., and D. M. Gordon. Colony age, neighborhood density and reproductive potential in harvester ants. *Oecologia* 119: 175–82.

1997. Gordon, D. M., and D. Wagner. Neighborhood density and reproductive potential in harvester ants. *Oecologia* 109: 556–60.

Page 82 Most food goes to larvae: 1985. MacKay, W. P. A comparison of the energy budgets of three species of *Pogonomyrmex* harvester ants (Hymenoptera: Formicidae). *Oecologia* 66: 484–94.

Pages 86–88 how foraging area changes year after year: 1995. Gordon, D. M. The development of an ant colony's foraging range. *Animal Behaviour* 49: 649–59.

Pages 86–88 on the cost of encounters between colonies. See the article cited for pages 75–77, above.

Chapter 5

Pages 97–98 on harvester ant response to oleic acid: 1958. Wilson, E. O., N. I. Durlach and L. M. Roth. Chemical releasers of necrophoric behavior in ants. *Psyche* 65: 108–14.

Ants treated with oleic acid "were carried off to the refuse, live and kicking": 1985. Wilson, E. O. In the queendom of ants: a brief autobiography. In D. Dewsbury, ed. *Leaders in the Study of Animal Behavior.* London: Associated University Presses, p. 473.

1983. Gordon, D. M. Dependence of necrophoric response to oleic acid on social context in the harvester ant, *Pogonomyrmex badius. J. Chemical Ecology* 9(1): 105–11.

Chapter 6

Page 106 Seeley on messenger bees: 1979. Seeley, T. D. Queen substance dispersal by messenger workers in honeybee colonies. *Behavioral Ecology and Sociobiology* 5:391–415.

Pages 106–7 How fire ants search an unfamiliar place: 1988. Gordon, D. M. Group-level exploration tactics in fire ants. *Behaviour* 104: 162–75.

Pages 107–110 how path shape, searching efficiency, and interaction rate are related: 1992. Adler, F. R., and D. M. Gordon. Information collection and spread by networks of patrolling ants. *American Naturalist* 40: 373–400.

Page 110 on how the Argentine ant disrupts the communities it invades: 1998. Human, K. G. and D. M. Gordon. Behavioral interactions of the invasive Argentine ant with native ant species. *Insectes Sociaux,* 46:159–163.

1998. Human, K. G., S. Weiss, A. Weiss, B. Sandler and D. M. Gordon. The effect of abiotic factors on the local distribution of the invasive Argentine ant (*Linepithema humile*) and native ant species. *Environmental Entomology* 27:822–33.

1997. Human, K. G., and D. M. Gordon. Effects of Argentine ants on invertebrate biodiversity in northern California. *Conservation Biology* 11:1242–48.

1996. Human, K. G., and D. M. Gordon. Exploitative and interference competition between the Argentine ant and native ant species. *Oecologia* 105:405–12.

Page 111 how Argentine ants adjust searching behavior according to group density: 1995. Gordon, D. M. The expandable network of ant exploration. *Animal Behaviour* 50: 995–1007.

1997. Gordon, D. M. Networking ants. *Natural History.* September 1997, p. 26.

Pages 112–13 comparing encounter patterns of three ant species: 1993. Gordon, D. M., R. E. H. Paul, and K. Thorpe. What is the function of encounter patterns in ant colonies? *Animal Behaviour* 45: 1083–1100.

Pages 114–16 how contact rate depends on group density in *L. fuliginosus.* See article cited for pages 112–13, above.

Chapter 7

For a review of current work on task allocation in social insects: 1996. Gordon, D. M. The organization of work in social insect colonies. *Nature* 380: 121–24.

Pages 119–24 perturbation experiments to study task allocation in red harvester ants: 1987. Gordon, D. M. Group-level dynamics in harvester ants: young colonies and the role of patrolling. *Animal Behaviour* 35: 833–43.

1986. Gordon, D. M. The dynamics of the daily round of the harvester ant colony. *Animal Behaviour* 34: 1402–19.

Pages 125–27 perturbation experiments with marked ants: 1989. Gordon, D. M. Dynamics of task switching in harvester ants. *Animal Behaviour* 38: 194–204.

Pages 127–28 Are reserves committed once they are recruited? See the 1989 article cited for pages 125–27, above.

Pages 130–34 how task allocation changes as a colony grows: See the 1987 article cited for pages 119–24, above.

Pages 134–37 The discussion of caste here refers to a research program most notably presented in Oster, G., and E. O. Wilson, 1978, *Caste and Ecology in the Social Insects.* Princeton University Press.

A more detailed version of my discussion here is in: 1989. Gordon, D. M. Caste and change in social insects. In *Oxford Surveys in Evolutionary Biology,* vol. 6, P. Harvey and L. Partridge, eds., p. 56–72. Oxford University Press.

Chapter 8

Page 143 on the dynamics of neural networks: 1982. Hopfield, J. J. K. Neural networks and physical systems with emergent collective computational abilities. Proceedings of the National Academy of Sciences U.S.A. 79: 2554.

Pages 143–44 a neural-network-type model of task allocation in red harvester ants: 1992. Gordon, D. M., B. Goodwin, and L. E. H. Trainor. A parallel distributed model of ant colony behaviour. *Journal of Theoretical Biology* 156: 293–307.

Pages 145–48 Model of task allocation: 1996. Pacala, S. W., D. M.

Gordon, and H. C. J. Godfray. Effects of social group size on information transfer and task allocation. *Evolutionary Ecology* 10: 127–65.

Page 149 Levins on models: R. Levins. 1968. *Evolution in Changing Environments.* Princeton University Press.

Page 150 An introduction to the self-organization models of social insect behavior developed by the group at the Université Libre de Bruxelles: 1986. Deneubourg, J. L., S. Aron, S. Goss, J. M. Pasteels, and G. Duernick. Random behaviour, amplification processes and number of participants: how they contribute to the foraging properties of ants. *Physica D* 22: 176–86.

Pages 160–61 Task is associated with recent encounter history: 1999. Gordon, D. M., and N. Mehdiabadi. Encounter rate and task allocation in harvester ants. *Behavioral Ecology and Sociobiology* 45: 370–77.

Pages 161–62 Ants of different task groups differ in the chemicals on their cuticles: 1998. Wagner, D., M. J. F. Brown, P. Broun, W. Cuevas, L. E. Moses, D. L. Chao, and D. M. Gordon. Task-related differences in the cuticular hydrocarbon composition of harvester ants, *Pogonomyrmex barbatus. J. Chemical Ecology* 24: 2021–37.

INDEX

•